T0008338

WIRED FOR LOVE

WIRED FOR LOVE

A Neuroscientist's Journey Through Romance, Loss, and the Essence of Human Connection

STEPHANIE CACIOPPO, PhD

FLATIRON BOOKS
NEW YORK

For information, address Flatiron Books, 120 Broadway, New York, NY 10271.

www.flatironbooks.com

Designed by Donna Sinisgalli Noetzel

Library of Congress Cataloging-in-Publication Data

Names: Cacioppo, Stephanie, 1974– author.
Title: Wired for love : a neuroscientist's journey through romance, loss, and the essence of human connection / Stephanie Cacioppo, Ph.D.
Description: First edition. | New York, NY : Flatiron Books, 2022. | Includes bibliographical references.
Identifiers: LCCN 2021047706 | ISBN 9781250790606 (hardcover) | ISBN 9781250790620 (ebook)
Subjects: LCSH: Neurosciences—Social aspects.
Classification: LCC RC343.3 .C33 2022 | DDC 616.80092—dc23
LC record available at https://lccn.loc.gov/2021047706

First Edition: 2022

The pages of this book are printed on a low CO_2 intensity uncoated groundwood paper. One hundred percent of the fiber stock required to make it came from sawmill residuals from trees certified by the Sustainable Forest Initiative and/or the Forest Stewardship Council.

10 9 8 7 6 5 4 3 2 1

TO

YOU

Contents

WIRED FOR LOVE

Introduction

You can't blame gravity for falling in love.

—ALBERT EINSTEIN

Paul Dirac was nobody's idea of Prince Charming. But he was a genius. In fact, after Einstein, Dirac was perhaps the most brilliant theoretical physicist of the twentieth century. He pioneered the field of quantum mechanics. He correctly predicted the existence of antimatter. In 1933, he won a Nobel Prize when he was just thirty-one years old. Yet, in terms of his personal life, the physicist was the social equivalent of a black hole. Colleagues described him as almost pathologically reticent and jokingly invented a unit called "the dirac" to measure his rate of conversation: one word per hour. At the University of Bristol, and then in graduate school in Cambridge, Dirac formed no close friendships, to say nothing of romantic relationships. He cared only for his work and was amazed that other physicists spent precious time reading poetry, which he thought was "incompatible" with science. Once, attending a dance with his fellow physicist Werner Heisenberg, Dirac looked out on a sea of swaying bodies and couldn't understand the point of this strange ritual.

"Why do you dance?" Dirac asked his colleague.

"When there are nice girls it is a pleasure to dance," he replied.

Dirac reflected on this answer for a long time, then posed another question.

"Heisenberg, how do you know *beforehand* that the girls are nice?"

In 1934, Dirac was introduced to a middle-aged Hungarian woman named Margit Wigner. Everyone called her Manci. She was his opposite in many ways: scientifically illiterate, extroverted, *fun*. But she took a strange interest in this aloof physicist. She detected a capacity in him that he did not see in himself. She wrote him love letters; he responded with shrugs, correcting her English and criticizing her appearance. She said he deserved a second Nobel prize—"in cruelty."

Yet she did not give up on him. She convinced him to spend time with her, to share his dreams, to confess his fears. He began, by degrees, to soften. When they parted after one long visit, he was astonished by an entirely new sensation. "I miss you," he said. "I do not understand why this should be, as I do not usually miss people when I leave them."

Dirac and Manci would eventually marry and spend a half century happily in love. In one of his letters, Dirac told his wife that she had taught him something that, for all his genius, he never could have figured out on his own. "Manci, my darling . . . You have made a wonderful alteration in my life. You have made me human."

All the Single People

Dirac's story illustrates how the power of love helps us realize our innate human potential. Understanding that power—why it evolved, how it functions, how it can be harnessed to strengthen our bodies and open our minds—is the subject of this book. It is a subject that has only gotten more complex in recent years. We live in a time when the environment needed to sustain love is being stressed in new ways. Marriage rates have plunged to historic lows. Half of adults in the United States are now single—compared to 22 percent in 1950. And while all these single people aren't necessary lonely—as we'll learn, there's an important difference between being alone and loneliness—those who are single not by choice but by circumstance are more likely to feel lonely. This includes many single parents. According to a nationally representative survey carried out in 2020, single parent households report higher levels of loneliness than other households; and a 2018 survey from Scotland showed that one in three single parents felt lonely frequently, while one in two reported feeling lonely "some of the time." Loneliness has in fact become so pervasive and so damaging that many public health experts describe it now as a full-blown epidemic, one that touches not only single people but also unhappy couples.

Maybe this yearning for human connection explains why online dating has been growing at an explosive rate. Between 2015 and 2020, dating app revenues jumped from $1.69 to $3.08 billion and are projected to nearly double again by 2025. And according to an online survey conducted in the last quarter of 2020, nearly 39 percent of single, widowed, or divorced internet users say they used an online dating service in the previous month.

Yet despite the sophisticated new algorithms designed to deliver the perfect partner—and some encouraging data about the success of long-term relationships formed online—many people report that dating has gotten more difficult in the last decade. While some of us find love, others keep swiping, looking for that special someone, feeling like the perfect match is within reach but not knowing how to connect.

Are we holding love to a higher standard than we once did? Is there something about digital-age dating that's fundamentally different from meeting someone in real life? Does the dating pool seem shallow to you? Or, on the flip side, are there just too many fish in the sea? The more you trawl, the more you worry that something is wrong with your net. While the common view is the more choice the better, research has challenged this idea by showing that people prefer a limited range of choices—often between eight to fifteen—over a more extensive array of options. Beyond fifteen options, people start to feel overwhelmed. Psychologists call this problem *choice overload*. I prefer the term "FOBO": fear of a better option.

Whatever you call it, it's exhausting—so much so that, for many single people on the dating market, the arrival of the Covid-19 pandemic gave them the excuse they were waiting for to close up shop and retreat into the safety of celibacy. As the pandemic started to ease, some single people actually started to experience FODA: the fear of dating again. Maybe they had been traumatized by the alienation caused by commodifying their self-worth and packaging it for consumption in a digital marketplace. Maybe they were *ghosted* one too many times. Maybe they were tired of looking for love and coming up short.

Of course, that wasn't everyone's story. While some people put their romantic plans on pause during the pandemic, the use of dating apps overall actually went up, as people sought connections online. And while many people were reluctant to date again after emerging from lockdown, other singles felt a jolt of energy, hoping to finally find the One by entirely changing their dating MO: Some turned to *type-casting* (only meeting prospective mates who "checked all the boxes"); others to *apocalypsing* (treating your next relationship like it's your last).

The pandemic wasn't only an enormous test for single people battling the effects of social isolation but also for those in relationships—who spent more time together than ever before. As happened during other global crises (the Great Depression, World War II), marriage rates fell, crashing through prepandemic lows. With their plans on hold, couples slowed down and got to know each other better—for better or worse. A Cambridge doctoral student in mathematics calculated that the average relationship aged by four years during lockdown. Some people wanted out; cultural commentators speculated that relationships on the rocks would not survive the stress of lockdown; and there were reports in the press about divorce lawyers being inundated with calls. But according to a survey conducted a few months into the pandemic, half of American couples said the experience of confinement actually made their relationship stronger; only one percent said they were worse off as a couple.

While the pandemic showed how resilient our relationships can be, there are still many challenges now facing couples. For all its social benefits, the rise of digital technology can be a mixed blessing for relationships. It all depends on how you use it. On the one hand, it can help people stay connected even when there's physical distance between them. On the other hand, the devices we carry around to connect with others can sometimes prevent us from connecting with our partner—even when they're in the same room with us. Two-thirds of people aged thirty to forty-nine say their partner is sometimes distracted by their phone when they're trying to talk to them. Thirty-four percent of people between the ages of eighteen to twenty-nine who are in relationships say that their partner's social media use has made them feel jealous or unsure about their relationship.

Adding to these new challenges are all the classic headwinds facing couples, like power struggles, lack of loving feelings, lack of communication, and unrealistic expectations, which according to relationship therapists are among the top reasons why couples split up.

All these challenges have driven many people to the brink of giving up on love altogether. A stunning half of single American adults—and a majority of single women—now say they are not even on the dating market, according to Pew Research. Across the globe, according to research from the United Nations, single living is on the rise and people are struggling to find suitable partners. Japan is a particularly stark case. About half the people there who want to marry say they can't find a spouse.

Many of these relationship trends seem to hit millennials the hardest. In America, 61 percent of them are currently living without a spouse or a partner. And while millennials may be struggling to find love, some of the youngest people who could be in the dating pool are actively avoiding it. A clinical psychologist who teaches a popular course at Northwestern University called Marriage 101 told the *Atlantic* that many of her students were steering clear of romance altogether. "Over and over, my undergraduates tell me they try hard not to fall in love during college, imagining that it would mess up their plans."

This Thing Called Love

I am not only a neuroscientist of love but also a hopeless romantic. And I am here to make the case that, in this time of social flux, when more of us are choosing to live alone and are tempted to turn away from romantic relationships, we should take heart. The world is changing, yes, but love will change with it. Love will evolve. This is one of love's best features, its adaptability. Yet while love is endlessly customizable, we must remember that it is never expendable. Love is not optional. It is not something we can do without. Love is a biological necessity.

My scientific research on the brain has convinced me that a healthy love life is as necessary to a person's well-being as nutritious food, exercise, or clean water. Evolution has sculpted our brains and bodies specifically to build and benefit from lasting romantic connections. When those connections are frayed or ruptured, the consequences to our mental and physical health are devastating. My research has revealed that not only are we wired for love but that, like Dirac, we cannot realize our full potential as human beings without it. Whatever the future of our social life will look like, love must be the foundation and cornerstone. While I discovered this in the laboratory by spending hundreds of hours scanning and analyzing the brains of those in love (as well as the heartbroken), I did not fully understand the importance and true beauty of love until I found, lost, and rediscovered it in my own life.

In this book, I hope we can unlock the mysteries of love together, but before we begin, we have to determine what it is we are actually talking about when we say that four-letter word. Although I will discuss other types of love in this book (maternal love, unconditional love, the love we feel for friends, pets, work, sports, our purpose in life), I am chiefly interested in *romantic* love, the kind of invisible bond that binds two human beings tightly together by choice alone, the kind that makes your heart go bumpity bump, the kind that launches a thousand ships, builds families, breaks hearts (quite literally, as we'll discover).

My discipline, which is social neuroscience, takes a holistic view of love. By looking deep into the brains of people in love, we discover that this complex neurobiological phenomenon activates not just the brain's mammalian pleasure centers but also our cognitive system, the most evolved, intellectual parts of the brain that we use to acquire knowledge and make sense of the world around us.

Yet people rarely look to neuroscience to help them understand something as majestic, as mysterious, as profound, as love. More often we turn to poets. For the length of a line of verse, someone like Elizabeth Barrett Browning can grab hold of that ineffable feeling called love. Let her count the ways: "I love thee with the breath, smiles, tears, of all my life." Maya Angelou elegantly describes all of us searching for love as "exiles from delight," people "coiled in shells of loneliness"—waiting for love "to liberate us into life."

But when it comes to defining love, poets can be, well, poetic. Take the French poet and novelist Victor Hugo, for example. Instead of answering the question "What is love?" he just dodges it with literary razzle-dazzle. "I have met in the streets a very poor young man who was in love. His hat was old, his coat worn, the water passed through his shoes and the stars through his soul." Or how about this plum, from James Joyce's *Ulysses*? "Love loves to love love."

As sentences, these are intriguing. As definitions, they are at best incomplete. Scientists must be precise, almost surgical, in our approach. To study love, we must dissect it. We must determine not only what love *is* but what it is *not. Is it an emotion or a cognition? Is it a primal urge or a social construct? Is it a natural high or a dangerous drug?*

Sometimes, as we'll discover, the answer is "both." Sometimes it's "neither." When hard-and-fast determinations are impossible, good scientists just continue to peel the onion.

A scientist must not only define her terms but she must also establish *boundary conditions*, circumstances under which her definition of love no longer applies. *Is love still love if it's not mutually felt? Is love still love in the absence of lust? Can you truly be in love with two people at one time?* Once we have clear boundaries for determining a solid definition of what love is, we can begin to investigate how this thing really operates, and even test whether some of the oldest saws about love have any scientific validity: *Is love really blind? Can you fall in love at first sight? Is it actually better to have loved and lost than never to have loved at all?*

By putting love under the microscope, we begin to generate (and answer) new questions that we would never have even thought to ask. *Why do people in love feel less pain? Why do they bounce back more easily from disease? Why are they more creative at certain tasks? Why are they better able to read body language or anticipate the actions of other people?* But just as we can evaluate the benefits of love, we can also examine the risks and dangers it poses. *Why do people fall out of love? Why does it hurt so much to lose love? How can you fix a broken heart?*

In this book, using my own research and that of my peers in disciplines from sociology to anthropology to economics, I will share with you what modern science has to say about one of the oldest facets of humanity. I will examine matters of the heart by looking deep into the brain. I will also offer a few case histories, drawn from my patients, my family, couples I've encountered, people who illustrate some powerful feature of how love works.

But the primary case study in this book is my own. Sharing that story, to some extent, goes against my nature. I am a shy and private person. Some of the things I've written down in this book will probably come as news even to my closest friends. For a long time, my only true love was science, and I assumed I would never experience romance outside the laboratory. Like Dirac, I found love unexpectedly—at first I was confused by it, but then I could not live without it.

When I was thirty-seven, in a flash of serendipity, I met the great love of my life. We dated across an ocean, got married in Paris, and, like two lovebirds, became absolutely inseparable. We traveled together, we worked together, we ran together, we even shoe-shopped together. If you put our seven years of marriage on the time clock of normal couples—who typically spend about six waking hours per day together—our union felt like the equivalent of twenty-one years. We loved every minute of it. We didn't feel time go by—we were too happy together—until the clock stopped.

I used to see love only through the lens of science, but my husband taught me to see it through the lens of humanity as well. And once I did, my life and my research were changed forever. So in this book, I have tried to tell both the story of my science and the science behind my story, in the hope that it will help you not only appreciate the nature of human connection but also give you some inspiration for how to find and sustain love in your own life.

1

The Social Brain

It was written in the skies

That the heart and not the eyes shall see.

—AS SUNG BY ELLA FITZGERALD

What happens when you take a typical wedding vow and rewrite it to reflect scientific reality? *Darling, from this day forward, I promise to love you with all my brain.* In making these words anatomically correct, we have robbed them of romance. The romantic version, the real version, the thing any bride or bridegroom knows to say while clasping hands with their beloved is *I promise to love you with all my heart*.

The heart is the organ we talk about when we talk about love—*not* the brain. To reverse these two is to translate the language of love ("you stole my heart") into something absurd, almost grotesque ("you stole my brain"). Today we know that the brain is chiefly responsible for emotions and cognition and ultimately for our ability to fall and remain in love. So why does our language still not reflect this reality? Why is it that we treat romance and passion as *matters of the heart*?

I believe that to truly understand love the first thing we must do is relocate it from the place where it has dwelled for most of human history. By that, I mean we must break the ancient bond between love and the heart.

This is no easy task. The *Oxford English Dictionary*'s entry on "heart" contains an impressive fifteen thousand words, most of them examples of how the term is used to describe love or other kinds of emotions, feelings, and thought processes. To lose someone we love is to be *heartbroken*. To revise an important decision is to experience a *change of heart*. To succumb to fear is to *lose heart*. To be kind is to *have a big heart*. I confess that despite my line of work, I use many of these expressions myself—maybe I'm a poet *at heart*?

These idioms aren't confined to English; versions exist in practically every other human language. And they go back at least as far as the twenty-fourth century BC, when the expression meaning "spreading wide his heart in joy" was carved inside an Egyptian pyramid. Similar expressions appear in the *Epic of Gilgamesh* (ca. 1800 BC) and Confucian texts (ca. 450 BC). Good luck finding such poetry about the brain in the ancient world.

What most people don't realize is that these expressions are not really metaphors. They are artifacts, dating from a time when everyone in the world, from Aristotle on, genuinely believed that our feelings originated not in our heads but in our chests. Historians of science have a fancy name for this belief: the *cardiocentric hypothesis*. And its origins are similar to those of geocentrism, the now-debunked idea that Earth is at the center of the universe and the sun and the planets rotate around it. Such a view may seem silly to us, now that we have telescopes and rocket ships, but in ancient times it conformed to what people experienced as their daily reality: The sun seemed to move in the sky while Earth, by all appearances, stayed put.

The same commonsense thinking led people to believe that our minds were in our chests. Just think about the feeling of being excited. Your heart pumps faster. Your breathing gets heavier. Your stomach tightens. And what does your brain do? As far as people could *feel*, it just sits there—inert, quiet.

In his search for the locus of the mind, Aristotle noticed that the loss of a heartbeat often accompanied near-death experiences. So Aristotle gave fundamental importance to the heart, blood, and blood vessels. In his cardiocentric view, the heart was responsible for thoughts and feelings. He also noticed that brains, unlike internal organs, were relatively cool to the touch. And so he deduced that our brain served as little more than physiological air-conditioning, tempering "the heat and seething of the heart," which he regarded as the true "source" of all our senses.

(Interestingly, recent research has shown that Aristotle was not entirely off base. Scientists have discovered that though our hearts might not control our brains, each organ constantly interacts with the other through hormones, electromagnetic fields, and even pressure waves.)

Even though Aristotle's cardiocentric view dominated in antiquity, there were others in his time and in the centuries that followed—such as the philosopher-scientists Erasistratus, Herophilus, and Galen—who believed that basic emotions, rational thinking, consciousness, and even mysterious phenomena like love originated not in our hearts but in our heads. Yet the exact role that the brain played in our anatomy remained an open question through the

Renaissance. As Shakespeare put it in *The Merchant of Venice*, "Tell me where is fancy bred. Or in the heart or in the head?"

Leonardo da Vinci also wondered about the mystery of the brain. According to Jonathan Pevsner, formerly a psychiatry professor at the Johns Hopkins School of Medicine who has published several papers on Leonardo's contributions to neuroscience, Leonardo saw the brain as the seat of the mind and the center of all our senses, a "black box" that receives, processes, and translates information. Around 1494, Leonardo drew three sketches hypothesizing the confluence of the senses—or what he called *senso comune* (common sense) within the brain ventricles. The ventricles are interconnected basins filled with cerebrospinal fluid that protect the brain from physical shocks, distribute nutrients, and remove waste. In his pursuit of knowledge, Leonardo achieved a perfect balance between art and science, and this was also true with his concept of the brain. He believed that visual information—"what you see"—was "processed in the main ventricle to help interpret the world." Leonardo explored other aspects of the brain, from blood supply to the cranial nerves. Although neuroscientists would later discover that brain matter—rather than ventricles—is the key to mental function, Leonardo's remarkable intuitive conjectures managed to expand the idea of the brain.

As the centuries progressed, Leonardo's vision was refined by a succession of pioneering investigators, who built the modern idea of the brain. Their names are revered in the history of neuroscience: Andreas Vesalius, Luigi Galvani, Paul Broca, and Santiago Ramón y Cajal, to mention just a few. Some dissected the brain to understand its constituent parts. Some injected ink stains into blood vessels to reveal connections between brain and body. Some made deductions about the function of different regions of the brain after examining patients who had suffered localized damage. These were the predecessors of modern neuroscientists: the predecessors of people like me.

A Magic Cabbage

For my neuroscience classes at the University of Chicago, I sometimes wheel into the lecture hall a glass jar in which a human brain floats gently in a bath of formaldehyde. I borrow this specimen from the neurobiology department, where a number of brains have been collected over the years, given to the university by generous donors who are enamored with science. Thanks to them, I can offer my students a unique opportunity to show them up close and personal—IRL (in real life), as they would say—the organ they study in such detail in their textbooks. I distribute latex gloves and ask: "Who wants to touch the brain?"

Ninety percent of my students raise their hands. The rest are content just to observe, or they've made arrangements with me beforehand to skip this class. Most students are dazzled by the opportunity to come in contact with the brain, to imagine this slippery organ inside their own heads, ruling their bodies and minds in a way that scientists like me are only beginning to fully understand.

But not everyone in the class is equally impressed.

"That's it?" one girl asks as I extend the brain out to her in my gloved hands. The smile on my face now turns bashful, like a waiter in a Michelin-star restaurant who has just theatrically lifted the cloche off a dish to reveal a tiny tomato. "I thought it would be . . . I don't know . . . somehow more impressive."

In a way, I can understand her disappointment. I've taught her that the brain is the most powerful and complex organ in the universe. And she is now faced with an object that, quite frankly, looks pathetic. It's a mess of fleshy pink and gray wrinkles, measuring about six inches long, weighing about three pounds, and, having been pickled in formaldehyde, has all the beauty of a boiled cabbage.

But let's slice this thing in half, separating the left brain from the right. What will we see inside? The wrinkly exterior gives way to a layer of smooth gray tissue. Known as *gray matter*, this part is richly concentrated with neurons, the nerve cells that are the brain's building blocks and are responsible for everything from information processing to movement to memory.

We have a lot of neurons—86 billion—yet it is not their sheer number that accounts for much of what we could call intelligence. In fact, as the distinguished neuroscientist Michael Gazzaniga points out, most of the neurons in the brain (about 69 billion) are found in the cerebellum, a small area at the base of the brain that coordinates our balance and motor control. The entire cerebral cortex, the part of our brain that is responsible for complex thinking and other aspects of human nature, contains "only" 17 billion neurons.

Much more important than the total number of neurons in our brain are the connections between our different brain regions. And connectivity is the specialty of the thicker region of nerve filaments packed deep inside our brain, beneath the blanket of gray matter. This is the *white matter*, the brain's information superhighway, which links different regions into powerful brain networks that shape both our conscious and unconscious experiences. In recent years, my fellow neuroscientists have identified and precisely mapped brain networks for all kinds of things, from motor skills to visual perception to language. I've made my own contribution in discovering the brain network responsible for the uniquely human experience of romantic love.

It's the volume and quality of those connective nerve fibers between brain cells—and not the size of our brains—that accounts for our incomparable skills as a species. And we have no shortage of them. In fact, if you unraveled all the white matter packed into the average twenty-year-old's brain, you would find that these microscopic wires extend more than a hundred thousand miles in length—or about four times as long as the circumference of the earth. Right now, in order to design artificial neural networks that many consider to be the future of computing, some of the best computer scientists in the world are studying how such a densely connected and economical biological system works. These scientists marvel at the brain's power and energy efficiency, how nature evolved a device that can store the equivalent of one million gigabytes of information—which equates to 4.7 billion books or three million hours of your favorite TV shows—while using the same energy as a single twelve-watt lightbulb.

Yet I believe that our neural wiring explains only part of the reason why our brains are so powerful. In addition to the vital connections inside our brain, we also depend on the invisible connections between our brains. By this, I mean our social life, our interactions not only with friends and loved ones but also our interactions with strangers, critics, and competitors. All this social activity, more than any single factor, has influenced the design and function of the brains we have today.

And like so many other stories in this book, the tortuous, mysterious, beautiful process by which our social nature sculpted the brains we have today is, at its core, a love story.

Love Made the Brain

The story begins in Africa, millions of years ago, with two of our earliest primate ancestors. Let's call them Ethan and Grace. Their romance began as a biological necessity. Yet once their relationship was consummated, Ethan and Grace decided to stay together. Grace had given birth to children that, compared to other mammals, were unusually helpless in their first few years of life. On top of figuring out how to protect them, the couple had to spend hours foraging to meet their dietary requirements. And then, to digest their raw food and store enough energy to live another day, they needed to sleep several hours each night. Juggling these tasks required social coordination. Suddenly, Ethan couldn't think only about himself; he had to see the world from Grace's point of view in order to anticipate what she needed.

Ethan and Grace had formed an intense preference for each other, a kind of relationship that biologists call a *pair bond*. Yet at some point in evolutionary history, their descendants—our human ancestors—took a giant leap, socially speaking. They adapted the skills used to build their own relationship (perspective-taking, planning, cooperation) and generalized them, forming bonds with other primates who were neither their reproductive partners nor their offspring. In other words, they made friends.

And these early humans needed friends because they occupied a vulnerable position in the food chain. They couldn't fly. They had no camouflage or armor. They lacked the strength, speed, and stealth of other species in the animal kingdom. They spent most of their time scavenging for food and evading predators. All they had, really, was an unusual talent for connection, a special knack for navigating the most complex environment in nature: the social world.

This was quite the superpower—and in the intervening eons, as anthropoid primates evolved, it would prove more decisive than their opposable thumbs, their skills at making tools, or the fact that they walked upright. As war and climate change made life on earth harsher, some species had trouble surviving; but these difficulties actually played to the strengths that early humans were developing.

Their social skills helped them build complex groups and eventually whole societies undergirded by mutual aid. People learned how to sort out friends from foes; to avoid predators; to anticipate the actions of neighbors; to privilege long-term interests over short-term desires; to use language to communicate; to manage mating relationships that were shaped not only by the female's ovulatory cycle but by different factors like affection and empathy. Finally, they learned how to trust and say "I love you."

According to the social brain hypothesis proposed by the British anthropologist Robin Dunbar in the 1990s, all these social complexities drove evolutionary changes in the brain that made us smarter. While humans started out with brains that were barely bigger than those of chimpanzees, our neocortex began to grow along with our social skills. Areas for language and abstract thinking blossomed. These higher-order regions did not only grow in size; they also became better connected to other parts of the brain. We can see the legacy of these changes by comparing the number of wrinkles (what neuroscientists call *convolutions*) in human brains versus those of less sophisticated primates, like baboons, whose brains are smoother and have fewer folds.

About seventy thousand years ago, the distant descendants of Ethan and Grace, our own species, *Homo sapiens*, moved from East Africa to the Arabian Peninsula and Eurasia. There they met other hominids, most famously the Neanderthals. The Neanderthals were fearsome competition: bigger, stronger, with better vision and brains that may have been slightly larger than those of humans. But the neural architecture of the Neanderthals and *Homo sapiens* differed in important ways. The Neanderthals had more space dedicated to vision and motor skills—they were ideal physical warriors. But the *Homo sapiens* were ideal social warriors: they could understand the intentions of others, they could consider a choice from two sides, they learned quickly from their mistakes.

All this allowed them to compensate for their shortcomings in strength. And, as a result, the epic evolutionary matchup between the Neanderthals and the *Homo sapiens* wasn't even close. By 11,000 BC, ours was the sole remaining human species. In other words, it was the need to interact with other people—first our significant others, then our friends, then the societies and civilizations that we built—that made us who we are today. And that process began with couples like Ethan and Grace falling in love.

A Neuroscience for a Social Species

Social connections have not only shaped the human brain throughout its evolution; they also continue to shape the brain throughout the course of an individual human's life. This is a fact that bears repeating because it's not at all obvious. After all, how many of us grew up thinking of socializing as expanding our minds? Rather we probably thought of it as downtime, something that we did as a break from our studies or creative pursuits, something that was not really important to our intellectual development.

Imagine how different our teenage arguments with our parents could have been if we had been armed with the latest insights from the emerging field of social neuroscience. "Actually, Mom, I don't have to get off the phone. Studies show that, by building and maintaining salutary social connections, I can *literally* grow my brain and am better able to focus on cognitively challenging tasks, like school. So Mom, pleeeease! Butt out!"

While it sounds far-fetched, this teenager's argument is valid. Neuroimaging studies do show that the size of core regions of the brain like the amygdala and the frontal and temporal lobes correlates with the size of our individual social networks. Similar findings reinforcing the value of social connections appear in studies of social species across the animal kingdom. If you raise a fish alone in an aquarium, its brain cells will be less complex than those of the same species of fish raised in a group. A desert locust's brain, when it's part of a swarm, grows by an impressive 30 percent, presumably to accommodate the additional information-processing demands of a more complicated social environment. Chimpanzees learn how to use new tools much faster when they are in groups than in isolation.

Yet just as my field reveals the benefits of the social world, it can also show us its dangers. Social pain—the heartache (oops, I mean brain ache!) that follows a bad breakup, for example—activates some of the same brain regions, like the anterior cingulate cortex, that respond to physical pain. People who report feelings of social isolation—what is typically called loneliness—have been shown to have less gray and white matter in key social areas of their brain. If they remain lonely, they are susceptible to a cascade of neurological events that echo throughout their bodies, leading to so many poor health outcomes that some public health experts now consider chronic loneliness on par with smoking as a grave risk to your health.

These are just a few of the insights to come out of social neuroscience, which studies how the connections between different individuals' brains—our social life—change what goes on inside our heads and our bodies. The discipline originated in the 1990s as a kind of surprising marriage between the so-called soft science of social psychology, in which the researcher must rely on observed external behavior and most likely subjective self-reports, and the so-called hard science of neuroscience, which uses high-tech scanners to peer inside the brain and precisely map its working parts.

Neuroscientists had previously treated the brain in isolation, thinking of it as a kind of solitary computing machine. This tendency to compare the brain to a mechanical device goes all the way back to the seventeenth century. French philosopher-scientist René Descartes, seeing water-powered automatons at work in the Royal Gardens in a suburb near Paris, thought that the human body worked in a similar way to these devices, that it was essentially a complex biological mechanism. And the Danish anatomist Nicolas Steno went even further, declaring that "the brain is a machine" like a clock or a windmill, and that the best way of understanding it was to take it apart and consider what the pieces "can do separately and together."

As the centuries passed, Steno's metaphor was updated. In the 1800s, the brain was compared to a telegraph system, sending and receiving signals to and from different parts of the body. In the second half of the twentieth century, it was likened to the personal computer: storing data in memory, processing information, executing commands. We social neuroscientists have further refined the metaphor. We see the brain not as a classic computer but as a smartphone with a wireless, broadband link to other devices. Just imagine how useful an iPhone would be without the ability to access the internet or send a text. Our brain also requires a strong connection to realize its full potential. And, like a smartphone, its connectivity makes it vulnerable. It can be hacked, cluttered with unnecessary apps, bombarded with distracting or anxiety-inducing notifications.

Yet the brain can also do something that smartphone designers can only dream of. It can reprogram itself. Neuroscientists call this *neuroplasticity*. And neuroplasticity is one of the true wonders of the mind. It refers to the capacity of the brain to grow while pruning inessential neurons when we're young; to expand and form new connections as we learn new things over the course of our life; and to repair or compensate for damage caused by an injury or the wear and tear of time. And social interaction is often the very thing that drives these vital changes inside the brain.

So, far from being a waste of time or incidental to the human experience, the connections we form with other people are quite literally the reason we exist as a species. Building healthy relationships also builds a healthier brain, one that—as we will discover—can stave off cognitive decline, spur creativity, and speed up our thinking. And there is perhaps no more powerful social activity, no better way of realizing our brain's full cognitive potential, than by being in love.

2
A Single Mind

The psyche is the greatest of all cosmic wonders.

—CARL JUNG

I sometimes think about what my teenage self would make of my life, what that solitary girl would have felt if she could have peered into a crystal ball and seen the future. I imagine her looking in on a crisp and sunny fall afternoon in Paris. The Luxembourg Gardens are alive with sounds—the rustling of chestnut leaves, the bells of the Church of Saint-Sulpice, the distant strains of an accordionist playing for tourists. A small group of people are standing on the grass. They are holding glasses of Champagne, listening to a man with salt-and-pepper hair and kind hazel eyes. He looks self-assured, like he knows he is making the right choice. And standing next to him, in a simple white dress, all grown up, is a shy yet confident woman with long hair who is holding a bouquet of white roses and saying "I do."

Wait, what? my younger self exclaims. *That's me?! I'm getting married!* This is the point in the fantasy where teenage Stephanie picks up the crystal ball and shakes it, thinking it must be defective.

I grew up believing it was my fate to be alone forever. And once I began to explore the neuroscience of love, I regarded my solitude as a kind of interesting irony, the sort of apparent contradiction that makes for good cocktail conversation. "Oh, look at me, the scientist who studies romance in the laboratory yet has never experienced it in real life." I told myself that being unattached made me a more objective researcher: I could investigate love without being under its spell. Unlike most people, I wouldn't take a preference for romance for granted. Rather, I would treat the pursuit of love as a fascinating, even mysterious, phenomenon in need of explanation. The fact that I was independent provided me this critical distance, not to mention all the uninterrupted time necessary to do research without distracting calls from boyfriends. So I tried to wear my solitude less like a burden than like a badge of honor.

And that, come to think of it, is what I have done for as long as I can remember.

I was born in a small ski resort in the French Alps and grew up in an even smaller town near the Swiss border. It looked like one of those snow globe villages. There was an old church, a bakery, a library, a primary school, cozy houses, and a ruined castle on a hill where I stargazed at night. I remember lying down on the grass surrounded by trees, and looking up beyond the vine-choked eaves of the castle into the night sky, searching for constellations. I sometimes hesitated to make social connections with kids at school, but I was mesmerized by the invisible connections I found between the stars, which turned the heavens into one giant celestial mural. For a time, I thought I wanted to be an astronaut. Maybe that's why I eventually decided to study the brain, because here was a universe just as vast.

If I were to play you the soundtrack of my childhood, there would mostly be silence—interrupted every now and then by my voice whispering a wish at the sight of a shooting star. At home, the music was different—the hiss of salted butter skittering across a crêpe pan. My family was half French (my father's side), half Italian (my mother's), and endlessly passionate about food. In the morning we dunked our croissants into big bowls of chocolate milk or café au lait. In the evening we had my grandmother's ravioli or linguine with Bolognese sauce. Food was family, and family was everything, all the social life I had growing up. When my relatives visited, we would spend our time talking about food, or preparing food, or eating food, or taking postprandial walks in which we discussed . . . what we'd eat tomorrow. It was a simple, happy, sheltered childhood.

Yet, from an early age, I noticed that there was something separating me and my cousins, who all had siblings and seemed more comfortable with themselves and with other children. They were happy with one another. But as an only child I had to learn how to be happy by myself. I remember spending hours listening to them play in the house as I stood outside, stargazing or hitting tennis balls against the garage in the moonlight. Although I always wished I had a partner to return the ball—and make the game more interesting—I found a kind of solitary bliss in the rhythm of the ball striking the wall. In the eyes of my cousins, I was the odd kid out. They described me as the dreamer, the thinker, and sometimes even thought there was something wrong with me because I spent so much time alone.

Children learn by watching others. And, when there are no peers or siblings around, sometimes the brain might invent them.

Research shows that while two out of three children will by age seven create an imaginary friend with whom they can be allies and share things, only children are particularly adept at this. And such imaginative leaps might explain a recent study showing that children without siblings tend to excel at "flexible thinking" and have more gray matter in the supramarginal gyrus, a cortical area associated with creativity and imagination.

While being an only child may be an advantage when it comes to thinking outside the box, there appear to be some trade-offs in terms of social brain development. Only children seem to have relatively less gray matter in a part of the prefrontal cortex that could help them resist temptation, delay gratification, and process emotional information, including the ability to infer the feelings of others. In some kids, this difference might produce a lack of interest in the social world. In me, it seemed to have the opposite effect. I was fascinated by the social mystery. I became intensely *interested* in social life, but rarely as a participant, mostly as the outsider looking in. I didn't understand why I felt different, why I didn't belong, why I couldn't blend into the social landscape. The older I got, the more curious I became about what I was missing.

There was just one person who made me forget myself, who made me feel like I was not an observer, like I truly belonged, a person with whom I could just *be*: my Italian grandmother, my mother's mother. Her name was Jacinta, but I called her Mémé. We spent hours together in her modest kitchen. She taught me how to make gnocchi by hand. She taught me the funny-sounding Italian dialect she spoke as a girl growing up in a small village between Venice and Udine, in the north of Italy.

She was obsessed with health. She instructed me to do a set of physical exercises, a kind of Old Country calisthenics, every morning that began with lying in bed with your legs straight up in the air and rotating each ankle thirty times. To this day, after I wake up, I sometimes kick my legs up and think of Mémé with a smile.

She was born in 1911—an eternity ago, from my perspective—and she believed that her morning routine was the secret to a long and vigorous life. She was an old-fashioned woman and a steadfast observer of tradition. An immigrant, she never had fine things, but she took painstaking care of her appearance: the ironed silk blouse, the carefully darned wool socks, the ancient string of pearls.

Her perspective on life was different from most members of my family. She had lost her first daughter to complications of appendicitis when she was just four years old. Afterward, my grandmother was never the same. She told me that she dressed up every day because she never knew which would be her last. Such care about her appearance did not extend to makeup. Her face was like a sheer cliff, weathered and beautiful. And her skin care regimen consisted of nothing but water and a gentle soap from the south of France made of plants and olive oil.

She said she was purified by church, which she attended every Sunday. She married young, like all the women in her village, and believed she had entered into a sacred covenant. For her, love was a gift from God, and if it ever felt like a burden, it was one to be borne with grace. Love for family, meanwhile, was unconditional—and she always managed to give more to those in need.

I was the apple of her eye, the solitary little girl who was always asking questions, who needed extra protecting, extra love. She doted on me. She spoiled me. She wanted to give me everything. When she wasn't visiting our house, she would call every night, like clockwork, to make sure I was healthy, happy, warm, tucked in. One night when I was nine, she rang and and told my mom that in the morning she would take the bus into a nearby city from the small town where she lived, to buy me a thermal sweatshirt that had just come on the market. She was concerned that I would catch my death playing tennis outside in the cold. I overheard this conversation and, for some reason that is still treated as a mystery in my family, I started crying, wailing uncontrollably, begging for her not to go. I screamed that I didn't need a sweatshirt—I just needed her. My parents were dumbfounded. I was not given to tantrums. *What had gotten into Stephanie?*

The next day, my father came to school to pick me up early. I remember the silent ride back home. We walked into the house, he sat me down, took a deep breath, and said, *"Mémé est partie"* (Mémé is gone). A stroke, a blood vessel bursting in the brain, had taken her away. It happened as she took her first step to climb aboard the bus. I found out later that Mémé's mother also had died from the same kind of stroke. I wonder how a child even makes sense of an event like this. Now I see strokes in neurological patients all the time, in high-intensity detail. I can picture the blockage caused by an ischemic stroke or the rupture from a hemorrhagic stroke. I know the millimeter differences between the kind you can recover from, those that leave you permanently disabled, and those that take your life.

But my nine-year-old mind couldn't possibly understand this, and the impenetrability of what happened made it all the more mysterious and terrifying. I think that Mémé's stroke, in a way, shaped my entire career, my whole view of the world. For years, I felt in my bones that I would meet the same fate as Mémé.

And I wanted to understand what had happened to her so I could somehow escape it, and help others do the same. From a very young age, I felt that I had a purpose in life.

But more immediately, losing Mémé underlined the smallness of my social world. And it deepened my sense of confusion, because now there was a new social ritual to deal with: grief. When I was nine, there was no class I could take about the meaning of the word "grief"—no book I could read at school about how to cope with it. I looked at my parents. They tried to stay composed, to conceal their feelings, so I would not be overwhelmed by sadness. But without an example from them, without siblings or friends to turn to, I had no clue how to deal with this torrent of new emotions. I cried every evening alone in my room until the day of the funeral.

At the church during the funeral service, I remember feeling lost. I wanted to pay my respects, wanted the world to know how much I loved my grandmother. I saw that most of the adults, sitting in the pews, had impassive faces, while one of my cousins cried her heart out. I thought, "Should I be crying more? Less?" Even in that moment I was perplexed by others, wondering about a protocol that to them probably seemed intuitive, instinctual.

That feeling of being the outsider, the spectator, dogged me throughout my school years. I eventually made friends, mostly with boys or sporty girls. I was skinny, with long, straight hair, scuffed knees, a tennis headband, and colorful sneakers. I liked running around the soccer field or building treehouses in the woods. I liked moving fast, feeling the cool alpine air and sweat on my skin. I didn't play only tennis. I skied in the Alps and swam in mountain lakes. I rode horses, ran on the track-and-field team, and played soccer. Yet when the action stopped, and the other kids began to talk, I clammed up, waiting for an opening in the conversation that I felt never came. I still don't quite understand exactly why social life was such a mystery to me. But I knew for sure that the biggest mystery of all was playing out at home.

My First Case Study

In 1942, a young commercial illustrator named Raymond Peynet was hanging around a pretty little park in the southern French town of Valence, when he witnessed something profound. On a wrought-iron bandstand, a lone violinist was serenading a young woman seated on a nearby bench who stared back at the musician with rapt attention, bright eyes, and burning red cheeks. Peynet had a front-row view of what the French call a coup de foudre (or lightning strike) and what Americans know as "love at first sight." He took his pencil to a sketch pad and drew the young violinist with his bowler hat and a broom of dark hair seducing the sprightly young woman in a slinky dress and ponytail.

Perhaps imagining the romantic future they would share together, he titled the drawing *The Unfinished Symphony*, but a magazine editor changed the name to *Les Amoureux*. And a legend was born. Over the next two decades, Peynet's lovers appeared everywhere: printed on scarves, postage stamps, wedding cards, porcelain plates, ads for Air France and Galeries Lafayette. Their love felt sweet without being saccharine, old-fashioned yet slightly surreal, and, above all, very French. As they snuggled under an umbrella in the rain or canoodled away the day on a Parisian bench, they looked like ideal avatars for a country that specialized in exporting *l'amour*.

I heard a lot about "*Les Amoureux* de Peynet" growing up because this was the nickname given to my parents by their closest friends. They thought my mom and dad—who never fought, who always held hands, who spent hours staring dreamily into each other's eyes—were the protagonists of a romance so charmed that it seemed to come straight out of a fairy tale. And just like Peynet's couple, my parents did experience love at first sight after a chance meeting, only it was not a violin that seduced my mom but a border collie named Marcel.

This was back in the early 1970s. My mother saw my father for the first time walking his dog in the park. Their eyes met and they smiled. They might have just passed each other by if it weren't for

Marcel, who yanked my father in her direction, panting with excitement and furiously licking her ankles.

"Marcel, no!" My father apologized. "He's not usually like this."

"It's OK, really!" my mother said, scratching the dog's ears, her smile getting wider. They began talking; they moved closer to each other. Marcel relaxed; his mission was accomplished. Before parting, my father asked for her phone number. He invited her out dancing the following Saturday. She must have liked his moves on the dance floor because she went home with him that night—although nothing happened except hours of conversation. My parents were cool in their own way—elegant, reedy, and, since this was 1970s France, given to wearing flared orange pants—but very conservative. And so they did not consummate their love affair until getting married, six months later. Perhaps that explains the quick courtship! Two years later, I was born.

As I discovered, love is an exceedingly complex phenomenon that affects the brain in many profound and mysterious ways. But *attraction*, the pull we feel toward someone we desire, is a simpler process, and we have a very good grasp on the biology and chemistry behind it. The first thing to note about attraction is that it's blazingly fast, which is not to say impulsive. We can assess the suitability of a prospective mate in less than two hundred milliseconds after seeing them. And when we're not interested—when we swipe left in Tinder-speak—our minds are made up even faster than that. This split-second decision is based on a host of complex visual information, deeply ingrained genetic preferences for reproductive fitness, as well as personal preferences of the "I like your style" variety.

Interestingly, one of the things we are often attracted to in a prospective mate is a vision of ourselves, although we never realize this. In one study, photographs of participants were photoshopped onto the faces of opposite genders—and not only did the men fail to recognize themselves as women (and vice versa), but they also rated their own photo as the most attractive. This tendency to fall for lookalikes might have applied to my parents, who look enough alike to be confused for brother and sister.

Smell also plays a critical role in attraction—but in an opposite direction. We tend to be attracted to people who have scents that are different from our own. Our nose picks up pheromones, chemical signals that contain genetic information. In one rather pungent study, female college students sniffed T-shirts of male students and ended up preferring the scents of students with immune systems that were very different from their own—a choice that would help protect their offspring by allowing them to inherit more ways of fighting off disease. Other factors, from the weather outside when you meet a new date or how hungry you are beforehand, might also play a role in partner preference—and not only in humans. For instance, studies show that hunger affects the mating preferences of female spiders, with well-fed female spiders being the most receptive to mates.

When we are not merely attracted to someone but begin to feel like we're falling in love, the brain releases a cascade of neurotransmitters and chemicals that transform our mood and perception. The first thing you'll notice if you've ever fallen in love is how *good* it feels. "Euphoria" is often the way we describe this state. And when you look at the process that's chemically powering that feeling, it makes perfect sense. Falling in love sets off a whole array of biological fireworks, including triggering a heart-shaped region called the ventral tegmental area (VTA), which pumps dopamine into the brain's "feel good" reward circuitry, the same regions that are triggered when people eat delicious but addictive food or drink wine.

This is why being in love can feel like taking a drug, without the hangover. Your heart rate increases, your skin becomes warm, your cheeks flush, your pupils dilate, and your brain sends signals to your body to release glucose for extra energy. The brain is flooded with a joyous bath of dopamine, but this is not the only chemical at play when you fall in love. Your brain also increases the production of norepinephrine, a related hormone that has the effect of giving you tunnel vision, focusing you on this important moment, and skewing your perception of time. The elevated levels of norepinephrine (along with dopamine) are part of the reason why so many people feel like time is fluid and even flying during the first date with a new lover. Meanwhile, your levels of serotonin, a key hormone in regulating appetite and intrusive anxious thoughts, plummet when you're falling in love to similar levels as that of someone suffering from obsessive compulsive disorder. So you might find yourself eating irregularly or fixating on small details, worrying about making "the right move" or sending "the perfect text," and then replaying the date over and over again in your head.

Physical contact between you and the person you're falling for triggers another extremely important hormone in human attraction: oxytocin, a neuropeptide that is sometimes called the "bonding hormone" because of how it acts like glue between people and increases our feelings of empathy and trust. It also spikes when you look deep into your partner's eyes or cuddle. In other words, oxytocin is essential to forming relationships.

And a recent study from Harvard Medical School showed that rising oxytocin levels may also explain why some people feel like eating less when they are in love. When researchers gave male participants, ranging in weight, an oxytocin nasal spray just before a large meal, they ate less (than a control group of men who received a placebo) and indulged in fewer postprandial snacks, particularly chocolate cookies.

Party of One

My parents settled in the small town where I later grew up, located just outside the French city of Chambéry. My mother became a professor of economics at the local college. My father managed a booming frozen-food business. While *maman* extolled the joys of academia and the beauty of constant learning, *papa* extolled the nutritional benefits of frozen peas—lectures to which I listened gamely because he would bribe me with large containers of ice cream.

My parents worked hard. Their jobs kept them apart for most of the day. But when they got home every evening, they joined like two magnets. After reuniting, they kissed each other on the lips—not a French kiss but a sweet little peck that genuinely expressed romantic affection. From that moment, nothing could separate them until they left for work the next morning. Rather than split errands and housework—the shopping, the cooking, the laundry—they opted to do it all together. When they sat on the couch, they invariably entwined, her hand in his, his arm around her.

I liked to watch them cooking together in our open kitchen. They could make the peeling, slicing, and frying of French fries look as choreographed as a ballet. Much of their communication was nonverbal; they could anticipate each other's actions with impressive accuracy. And they took such pleasure in their time together. I remember my mother was once making her famous family recipe for Bolognese sauce from scratch, and at one point she and my father started convulsing with laughter. Like any Italian, my mom can't speak without using her hands and had been so absorbed in their conversation that she failed to notice when she inadvertently started flinging large quantities of tomato sauce into my father's face. A food fight ensued, which ended with a long kiss.

Growing up, I knew my parents fit together perfectly. But where did I fit in? They loved me, sure, but I often felt like an unwanted appendage, a tagalong. In the family car, I sat in the backseat, in the center position, and would crane my head between them, sometimes resting my chin atop their clasped hands on the armrest as we drove. When I was little and they used to leave me at Mémé's house so they could spend a weekend alone together, I accused them of "abandoning me." I spent not minutes but hours staring at the parking lot, hoping they would come back and include me in their plans. (When I think of this today, I'm reminded of the way my dog stares out the window, waiting for me to return from work.)

When I got a bit older, I doubted whether I could ever find someone to complete me the way my parents completed each other. I sat alone on the couch watching them coo in the kitchen as I listened on my Walkman to "La Solitudine" (Loneliness), a moody Italian ballad by Laura Pausini.

> *This silence inside me, it's the fear*
> *Of living life without you.*

Living life without someone seemed to be my peculiar fate. I'm not exactly sure why, even today. Was I afraid of falling short of the romantic ideal that my parents had set? Was I just a budding workaholic, so in love with school and sports that I didn't leave any time for romance? Had I become too enamored of the French saying "It's better to be alone than in bad company"? After I heard it once, it became my mantra, or at least a good excuse. Certainly, I knew how happy love could make people. And I even played matchmaker in school, applying my interest in social dynamics to read the body language of my friends, noticing who was interested in whom, suggesting they get together. "Psst, Rachel, do you see the way Jean blushes when you come near?"

But I wasn't interested in playing that game for myself. I hung out with athletes or science geeks, and if I got close to any of them I felt more like they were the brothers I never had rather than prospective boyfriends. Once in junior high a boy on the tennis court stole a kiss—my first—and then told me that he'd only go out with me if I cut my long hair. I just smiled and walked away, my long hair flying in the wind.

After a while I started getting self-conscious because people—mainly my mother—started asking *why* I seemed to have zero interest in love. And how I had made it through junior high and high school without a boyfriend. When I went to university to study psychology, I would never mention my love life (or lack thereof) to my mother. When I told her about my good grades or how I scored a competitive internship, she would say, "Yes, but have you found the One?"

She confessed that she prayed every night that this long-awaited One would reveal himself. And sometimes she tried to give destiny a push. When I was fourteen, a few kids from school organized what we called in the late 1980s a *boum*, or dance party, for a friend's birthday. At that event, my mom cozied up to a couple who thought their son (who was also fourteen) would be a great match for me when we both turned twenty-one. Yes, Mom always thought long-term! So the eager parents pushed us to dance together that afternoon, and when we did they were so happy that they took a picture.

When I look at the photo today, I can't help but laugh—the boy and I danced so far apart that you could fit a table between us. He would go on to close that gap with someone else. A decade after our dance, he got married, settled down, started a family. I remained, at twenty-four, defiantly, happily single. And my mom? She became more worried than ever.

Soon it was impossible to get through a Sunday dinner without her lamenting that she'd never have grandkids. I told her to be patient, to find a new hobby instead of obsessing over my personal life.

"Have you thought about getting a puppy?" I suggested.

As uncomfortable as I felt, I understood her anxiety. She relied so much on my father, and he on her. They both derived security and meaning from their marriage. They wanted that for me.

The truth is that I concealed from my mom—maybe because of how insistent she was—the few experiences I did have with dating. I went out for a couple weeks with a guy who, maybe for most people, would have checked all the boxes. He was rich, well connected, handsome, aristocratic—the kind of guy you could easily imagine in epaulets riding a horse. When we met, at a charity ball in Monaco a million miles from my comfort zone, my friends (who had dragged me there kicking and screaming) joked that I had found my Prince Charming. They giggled excitedly when he asked for my number.

On our first date, he took me to the French White House, the Elysée Palace, where we walked through formal gardens and drank Champagne in rooms dripping with gold leaf. Looking at his flowing blond hair and movie star smile, I sighed a little for my mom's sake—this was *exactly* the kind of person she had always hoped I would marry. Yet I couldn't wait to get out of there, back to my schoolwork, back to the tennis court.

Often, I wasn't so much avoiding romance as being oblivious of those who had feelings for me. Such social cluelessness isn't easy for a social neuroscientist to admit. Yet I was just being myself—the constant observer, never the protagonist, of my own life. In grad school, for instance, I shared an office with a friendly medical student. We collaborated on research, we theorized for long hours, we argued and teased each other, and we laughed a lot. A few years later he confessed that he was sending me signals—the kinds I could read so easily when they were directed at other people—but for some reason none of them landed.

I think, deep down, I was open to the idea of true love, that I could have fallen for the right person. I even had a mental picture of my ideal match—kind, athletic, intellectually inspiring. But I didn't want to spend my whole life searching for that special someone. I wanted love to find me, and when it did, I wanted love to feel unambiguous and natural. I wanted it to be like what my parents had. I wanted it to give me the sense of purpose I got from my studies. And I wanted it to deliver the dopamine hit I got from sports. As kind as he was, nothing I experienced with that rich guy came close to the ecstasy of a perfect backhand in tennis that bounced right off the line. I wanted love to feel that way, like hitting the sweet spot, a feeling that doesn't get old.

And if it didn't, well, then maybe it wasn't for me. Who says that it takes two to live a fulfilling life? What if it was only social pressure that made marriage the norm? What if one wasn't the loneliest number after all?

3
Passion for Work

Science is not only a disciple of reason but, also, one of romance and passion.

—STEPHEN HAWKING

I told you that I never fell in love when I was young, but the truth is there was someone. I can't remember his name, but I'll never forget his mischievous grin, his piercing amber-colored eyes, and the way his entire body, from head to toe, was covered in soft brown fur.

I'm talking about a monkey, a primate all of two feet tall called a macaque, who changed my life one day in the summer of 1999. I was twenty-four and, like a lot of people at that age, uncertain of what exact career path to take. I was doing graduate work in psychology but was increasingly interested in the biology of the brain, the hard science behind the mind. The more I researched, the more I wondered how one could hope to fully understand human nature without exploring the nature of the very organ that *makes* us human.

I volunteered to do a presentation on the brain for the benefit of the other students and spent weeks diving deep into the neuroscience literature, drafting a presentation that grew monstrous in its complexity. I was fascinated, exhilarated—for weeks I could talk about nothing else. On the day of the presentation, I came to class with a huge smile on my face and launched into my talk with the fervor of an evangelical preacher. And at the end, catching my breath, I looked up at my kindly old professor, who had encouraged me to do the presentation in the first place, and saw that he was sound asleep.

I couldn't believe it. I burst into tears in front of my classmates and ran out of the lecture hall. The professor later apologized—he was taking a new medication, which had made him drowsy. He felt so guilty that, to make it up to me, he arranged for me to visit the lab of a famous French neurophysiologist so I could learn firsthand how the brain worked.

"*In vivo*," he added with a meaningful glance.

At the time I had no idea what *in vivo* meant, but I said, "Sure, why not," and just hoped that this Professor "Invivo"—whoever he was—would not fall asleep on me as well.

I drove my little Renault two hours from my parents' home to the Lyon campus of France's foremost scientific research institute, CNRS. The laboratory was quiet and sterile, but I felt an exciting energy coursing through those walls, the feeling that a major breakthrough was just around the corner. And this was where I came face-to-face with the monkey that changed my life. The macaque was standing in a cage—a fact that still makes me flinch—but he nevertheless seemed happy to see me, blinking his adorable eyes and squealing with apparent delight.

In vivo, a grad student explained, actually meant investigating how the brain worked inside a living organism. In this case, the organism was a macaque who stood on a platform with a series of electrodes surgically inserted into his brain. In the early days of neuroscience, listening to live brain activity through implanted microelectrodes was standard practice. The louder the sound, the more intense the activity. This technique owed much to the pioneers of the field, such as the Italian physician Luigi Galvani, who discovered in the eighteenth century that nerve cells and muscles are electrically excitable, and the Germany physicist Hermann von Helmholtz, who figured out in the nineteenth century that the electrical current in a neuron was actually relaying a message.

I was not given too many details about the study that day because of confidentiality rules, but I knew that the neuroscientists Jean-René Duhamel and his wife, Angela Sirigu, were investigating the power of the ventral intraparietal zone, which is abbreviated VIP. (I will spare you the usual jokes I make to my students about the brain's "VIP room.") Located just above your ear in the parietal lobe, this area helps primates (including humans) have a sense of where they are going and process all the visual, tactile, auditory sensations that swirl around the body as we move through the world. This brain area plays a big part in directing our eye gaze to help us avoid bumping into objects when we walk or run. It also helps us turn our heads, the way we might do—or want to—when someone charming walks by.

Following the standard procedure, the researchers connected the intracranial microelectrodes—which track the live electrical activity from local signals in the brain—to an amplifier, so they could literally play the sound of neurons firing from the VIP as the monkey changed his eye gaze.

"Do you want to listen?" one of the researchers asked me.

I nodded my head yes—I was almost too excited to speak. As I placed the headphones on my ears, time seemed to slow down. I could feel my heart beating fast. The macaque's neurons sounded mostly like static, yet there was a strong signal in all that noise. It was as if I had tuned in to the best radio station in the world. I was overwhelmed by the purity, the authenticity of that information, by this basic emanation of life. I knew in that moment of pure happiness I had found my calling. It was love at first sound.

Brushstroke

As mesmerized as I was by the sound of a live brain in action, I knew I could never work with caged subjects. With all due respect to Dr. Duhamel and Dr. Sirigu, part of the reason I wanted to spend my life studying the brain was to set people free. I thought the most direct way to do that while furthering my interest was to help people recover from debilitating injuries and brain disorders like epilepsy. So I enrolled in one of the best neurology departments in Europe, at the Hôpitaux Universitaires de Genève (abbreviated as HUG—no pun intended) in Switzerland.

During my first years in the PhD program, I still lived with my parents just over the French border, but I spent hardly any time at home. I would wake up before dawn to catch a 6:00 A.M. train to Geneva and wouldn't get back until after midnight. The neurology floor became my new home. I was so excited by my work that I didn't need much sleep.

I soon felt like some kind of brain detective. My job was to discover, in the aftermath of stroke or epilepsy or any brain injury, what part of the brain was preserved and what part of the brain was dysfunctional and had to be rehabilitated, or, in the case of intractable epilepsy, what part of the brain could be removed by a neurosurgeon without causing any long-term behavioral or neuropsychological deficits. Each case was as fascinating as it was emotionally demanding. In time I learned how to be compassionate while also dissociating, how to maintain enough distance to do my job without falling to pieces when I met an athlete who had lost the ability to walk, or a mother who could no longer recognize her own children.

One of the first clues I had that love and passion played an unusually important role in the inner workings of the brain came through my experience examining a seventy-one-year-old Swiss patient named Huguette. She was a successful painter and textile designer who had achieved some renown as an art teacher in Geneva, where she even had her own television show for a time. Art was her life. It gave her not only a living but also a way of thinking and interacting with the world. She never left her home without a sketchbook. And she shared her passion with her beloved husband, who was a celebrated painter.

I first encountered Huguette in the hospital one day in October 2001. She looked haunted, having just lived through the most terrifying twenty-four hours of her life. She had awoken suddenly in the middle of the night and, unable to go back to sleep, had gone to the kitchen for a drink of water. Yet, as she walked down the stairs, a strange sensation overwhelmed her. She felt "disoriented" in her own home. She could not recognize the walls in her stairwell. When she stumbled into the kitchen, she could not locate the cabinet where she kept the cups. Was she sleepwalking? Was she having a bad dream? She pinched herself. She knew she was awake. What was happening to her?

She tried to ignore the sensation. She returned to bed and eventually fell asleep. In the morning, Huguette felt more or less normal, except for the headache—an awful, pounding headache. She took some pain medicine, tried to get on with her busy day of painting and teaching.

Around 1:00 P.M., she got in her car to pick up her husband at his own art studio, about two miles from her house in Geneva. But while driving through her neighborhood—a neighborhood she knew so well she could navigate it with her eyes closed—she realized she had no idea where she was or in what direction she was headed. She kept going around and around a traffic rotary, unsure of what exit to take. Then came a loud crumpling noise. She had smashed the left side of her car into the rotary. She slammed on the brakes, stumbled out of the car. She was disoriented.

A passerby rushed to her aid. "Madame, are you OK?"

"I don't know. I'm lost. I don't know where I am."

"Where do you live?"

She had no problem supplying her address. And, once home, she had no problem recognizing her family, who immediately called an ambulance. Huguette was rushed to the hospital. She was given a CT scan to check for a tumor or a hemorrhage and a surface electroencephalogram (EEG) to rule out an epileptic seizure. The tests showed that she had suffered a serious stroke in the right parietal lobe.

Sitting atop the cerebral cortex, the parietal lobe is a fascinating part of the brain that I have studied in great detail. Among many other functions, it helps us makes sense of what we see. The parietal lobe contains the VIP—the region I heard firing in the brain of the macaque—which directs our eye gaze and helps orient our spatial attention to our surroundings. The parietal lobe is also responsible for our body image (the way we see ourselves in our mind's eye) and for visual attention (what we choose to focus on, and what we choose to ignore).

Based on Huguette's symptoms and the location of the lesion, we knew that her stroke had caused a major attentional impairment. But we didn't yet know what kind. We checked her into the hospital for a series of behavioral tests. While we awaited the results, we had an unexpected diagnostic breakthrough.

It had to do with breakfast. Huguette was annoyed—her tray seemed to be missing half its contents. Why, she asked politely, was she not given orange juice and a bowl of fruit? The patient who shared her hospital room had these items on her tray. "Why are they missing from mine?"

A nurse looked down at Huguette's tray and tried to conceal her surprise. The orange juice and the fruit were right there, sitting on the left side of the tray, in plain view. Yet somehow Huguette could not see them. For her, they did not exist. A light went off in my head. I asked to see Huguette's sketchbook. I had noticed that she had been drawing since she arrived at the hospital. I thought this was a coping mechanism, a way of dealing with the uncertainty of her diagnosis.

I smiled when I looked at the drawings. There were sketches of the nurse, of the doctors at work, of a beautiful woman with a veil whose image Huguette found in a fashion magazine. The drawings were lovely—light, playful, full of life—yet there was something about them that was also, undeniably, strange. They contained startling omissions and distortions of a kind that Huguette—who I later discovered was an excellent draftswoman with a great attention to detail—seemed not to notice. And these omissions and distortions were confined in every case to the left side of the page. She drew people with missing left arms, missing left eyes. She drew a woman wearing only the right half of a blouse.

My diagnosis was swift. The patient was suffering from a left-sided *spatial hemineglect*, a kind of attentional blindness (or mind's eye blindness), which had the effect of obscuring half of her world.

Because the damage occurred in the right hemisphere of her brain—which controls the left side of the body—it was affecting only the left side of her mind's eye. Huguette's eyes could still technically *see* everything around her, starboard and port, but her mind's eye would only pay attention to objects on her right. It wasn't as though the left side was blacked out; rather it ceased to exist, to matter. The halved world appeared to her whole. Whether it was a glass of orange juice, or a car, or a duck floating on the left side of Lake Geneva, Huguette literally did not know what she was missing.

Short of actual blindness, what could be more devastating to an artist? And yet this was not her only problem. Huguette's self-perception also seemed to be affected by the stroke. When she looked at her left hand and foot, they appeared gigantic, as though she were now seeing them through a magnifying glass.

"Do you think I will ever be able to paint like I did before?" she asked me.

I tried to be reassuring. "I know you will."

But I also told her that she would not be able to regain her abilities without undergoing months of rehabilitation. We began immediately, yet it soon became clear that the normal rehab program was failing her. Stroke patients are typically given a series of exercises that look a bit like children's games. They can spend hours doing such simple tasks as putting wooden pegs into small holes. Huguette looked at these "toys" we brought with unvarnished contempt.

"How is *this* going to help me paint again?"

Her reaction was typical of high-functioning patients—CEOs, artists, athletes, engineers—who become depressed or frustrated when forced to do remedial tasks that seem far beneath their normal capacity. Part of Huguette's brain was damaged, sure, but her identity was completely intact. And so, any treatment regimen had to take this fact into account.

That's why, when Huguette flat out refused to do the textbook exercises, I decided to throw out the textbook. If she was only interested in art, fine, we would use her passion as the conduit to recover her mind's eye. I took a tough love approach. I told her she was now enrolled in a new art class, and I was the instructor.

What was the curriculum? Sketching out new connections in the brain. The goal was to use the brain's natural power to adapt to tap into preserved, healthy regions and create new links that would compensate for or work around the damage. In Huguette's case, that meant months of difficult rehabilitation sessions that focused on her passion, and her identity, as an artist. In short, she had to teach herself how to draw again, how to expand a canvas that had radically shrunk. It was difficult, painstaking work.

In the first three weeks together, she made over sixty drawings in an effort to recapture her vision. Despite the fatigue, despite the depression, she did everything I asked. She often felt like giving up, like disappearing into the big scarf she sometimes wore in the chilly hospital halls. But giving up meant losing her identity as an artist.

After the latest exercise, when it felt like we weren't getting anywhere, she would ask me, "What's the point of all this?"

I took a deep breath.

"Your parietal lobe," I explained, "is like a large house with many rooms. And one of them has gone dark. Blown fuse, short circuit, whatever the cause—the lights in that one room are now out. And you can't turn them back on. So how will you see? How can you paint in the dark? Well, you need to turn the lights on in every other room, to open all the doors, to knock down walls if necessary, to *flood* the house with so much light that it doesn't matter that there is an outage in one room. Because now the entire house is illuminated."

And that's exactly what we did. I asked Huguette to draw her self-portrait from different angles, using mirrors to reflect her image from the right side to the left. Over and over again, I manipulated her field of vision, forcing her to pay attention to the neglected side, to "visit" other "rooms" in the house of her mind, to switch on more lights and to break down more walls.

She began to make progress, first in small steps, later by leaps and bounds. I was struck by the fact that if she weren't an artist—if she hadn't had this passion, this love, for her work—she never could have recovered from such a massive stroke. Incidentally, love seemed related to many of Huguette's breakthroughs in rehab. When we placed a photograph of a beloved family member—a grandchild, say—on her neglected left side, she noticed it more easily and faster than an image of an object or anonymous person. The positive associations she had with her grandchild triggered a powerful emotional response that activated the brain's limbic system. This system manages emotions and memories and is particularly effective at picking up on signals and sending them to the parietal lobe, giving her the "push" she needed to overcome her attentional deficit.

Slowly, as she built new mental connections, formed new habits, found new ways of seeing, she learned how to push her attention onto the left side. First the objects there appeared to splinter into jagged strips when she focused on them, almost like the sections of a stained-glass window. But in time the picture became whole. The canvas unfurled.

After a year, she had made a nearly complete recovery. Not only that, but she also emerged from rehab with an even better understanding of angles and proportions, a keener sense of every brushstroke, a deeper understanding of her identity as an artist. She confessed to me that, before the stroke, she sometimes felt overly cautious in her work. She even had an inferiority complex toward her artist-husband, who was more commercially successful.

After the stroke and her recovery, all those insecurities disappeared. Her canvases became bigger, her style looser and more experimental. She even started projecting colored lights onto her works and starred in an exhibition at the hospital. "I can't believe I'm saying this," she told me, "but I almost feel that this stroke has set me free."

Plasticity Makes It Possible

Huguette's recovery illustrates the power of the brain to rewire itself, that extraordinary feature known as *neuroplasticity*, which is fundamental to understanding how this organ works.

Neuroscientists often talk about the *function* of certain brain regions. They say that a given region helps us store memories (like the hippocampus) or detect danger (like the amygdala). This can give you the false impression that complex behaviors are exclusive to one part of the brain or another. In fact, this is rarely the case. On the contrary, the brain prizes versatility and likes to spread out its work across many regions.

Language, for instance, is not localized in a single part of your brain. Many areas are involved in its production and processing, from the frontal lobe to the parietal lobe to parts of the temporal lobe and insular cortex, to name just a few. I will discuss these regions in detail in the following pages because they are also intimately involved in our ability to form loving relationships.

Each individual brain area also has multiple functions, and so different brain regions may complement, reinforce, or, when necessary, even duplicate each other's work. They also may adapt and develop a new function to pick up the slack when another region is damaged.

All this might seem strange if you're inclined to think of the brain as a machine, like a car, that's made up of discrete parts that perform specific tasks. In a car, if the air-conditioning breaks, you wouldn't expect the fuel injection system to start miraculously blowing cool air into your face.

Yet in the brain this kind of jury-rigging happens all the time. When there's damage, the brain will try to adapt to preserve functionality. It has many neural pathways, many ways of reaching the same destination. If a pathway is blocked, it can sometimes reroute a signal along another pathway. This tendency for the brain to compensate for damage also explains why losing one of our senses (such as sight) tends to heighten another sense (such as hearing). The brain is trying to make up in one place for a loss in another.

This is what happened to Huguette. Although she failed to notice things on her left, she became unusually sensitive to objects on the right side in the aftermath of her stroke. And when she finally recovered following rehab, she actually became a more analytical painter. Her brain not only compensated for the damage, but the new mental connections that she built also left her feeling that, in addition to recovering the functionality she had lost, she had gained a fascinating new perspective.

Although Huguette was unique in many ways, her recovery followed a pattern that I soon began seeing in scores of other patients. Often it was their passion for the thing they loved most in life—whether that was a vocation, or a hobby, or a person—that helped them rediscover the skill or ability that they had lost. I had read about the "power of love" in pop psychology books, heard it sung about in ballads, even marveled at it in my parents' kitchen growing up, but now I was learning that it might actually play an important and undiscovered role in the brain. And I began to wonder if love might be the key not only to helping an injured brain recover but also to helping a healthy brain thrive.

4
The Love Machine

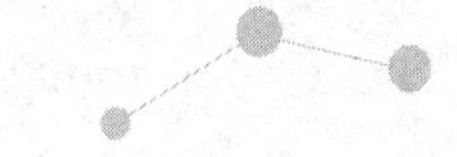

All truths are easy to understand once they are discovered;
the point is to discover them.

—GALILEO

Before I came along, only a few other researchers had tried to use the tools of neuroscience to study love. One reason for this is that it is an exceedingly difficult subject to tackle. The way the brain encoded the connection between two people was not something we could easily discover, much less measure or put into a mathematical equation. I felt a bit like Newton contemplating gravity, an invisible force that I knew existed but could not yet explain.

There was also another, more delicate, problem: skepticism from my fellow neuroscientists about whether exploring the neural basis of love was even a worthy subject in the first place.

"The neuroscience of love? Please tell me this is a joke," sneered one of my faculty advisers in Geneva. "It's career suicide. No one will fund you. No one will publish you."

To him, it was almost as if I were creating the scientific equivalent of cotton candy, as if anything related to romance wasn't sufficiently serious and substantive. He wouldn't be the last man to tell me that love was too fluffy a topic for a serious scientist, but he was probably the most direct. And because he got to me while I was still in grad school, he had the best chance of altering my course.

"You are working so hard to get your PhD. Why would you throw it away on a subject that's so common, so . . . simple?"

Simple? His description shocked me. The formula to make salt in chemistry is simple—one part sodium, one part chloride. The formula to make lasting love? That was way more complex. And broad-minded scientists understood that. I wish I had known then about the work of Peter Backus, an economist who had calculated that there are likely more intelligent alien civilizations in the universe than there are eligible women for him on planet Earth.

Love was anything but simple. As this adviser was talking, I was thinking of the many female social scientists who came before me, pioneers like Elaine Hatfield, Ellen Berscheid, Barbara Fredrickson, Lisa Diamond, and Susan Sprecher who had paved the way to study the psychology of love with rigor.

After politely listening to the professor's lecture, I excused myself, gently shut the door to his office, and then whispered under my breath: "Bite me." How could a self-respecting scientist dismiss something clearly significant to the human experience just because of how it *sounded*, or how it *looked*? Wasn't it our job as scientists to pose the questions that other people hadn't even thought to ask?

In his defense, there *was* a legitimate technical question about whether the thing that people called "love" was too vast, too unspecific, too subjective to study effectively. Was it just a compound of more basic feelings like attraction and attachment, a "secondhand emotion," as Tina Turner once sang? Maybe "love" meant entirely different things to different people based on personality, or class, or culture? Maybe I would be better off narrowing the scope of my research?

Such a view was neatly illustrated by a grant proposal that I submitted with the word "love" in the title. It was rejected. I later sent in the same proposal, nearly word for word, with one change: I substituted "love" with "pair bonding," and received the grant money.

While there may have been some initial hesitation among academics about the value of studying love, the popular press was very quick to embrace my work—especially around Valentine's Day, when I got interview requests from magazines like *Scientific American* and *National Geographic*. After the first few articles, colleagues began teasingly referring to me as "Dr. Love." The media coverage also attracted the attention of undergraduate students, who soon developed a personal stake in my research, thinking it could help them navigate their own budding campus romances.

By 2006, I had moved from Geneva to Dartmouth College in New Hampshire, where I was doing research in the department of psychological and brain sciences with the world-renowned neuroscientists Scott Grafton and Michael Gazzaniga. A new language, a new culture, a new climate—I was lost until I found my way into the laboratory where they kept the brain scanners and computers and I could spend my nights and weekends poring over data.

Very often, during office hours, a female student would show up at my door with a special request. Usually, she had a friend or two with her for support. She had heard about my work from the flyers I put up in the library: WANTED: WOMEN IN LOVE.

After a tentative knock on my office door, I heard some awkward throat clearing. "Excuse me, uhhh, Stephanie . . . do you have a minute for a chat?"

While I could write scientific papers in English, I was still learning the everyday language, and the word "chat" threw me, since in French it means "cat." But I got the sense that she wanted to talk.

"Please, have a seat."

The student sat down, dug her hands into her jean pockets, and began to blush. Her friends gave her a nudge.

"Go on, just ask!"

"OK!" she said, before turning to me to make a request I would hear over and over again that year: "Do you think I could use your Love Machine?"

On my patent application I actually called it a "System and Method for Detecting a Specific Cognitive-Emotional State in a Subject," but the students preferred "Love Machine," so that was the name that stuck. It was a ten-minute computer-based test that I designed, which, the students believed, could help them make up their minds when deciding between two potential romantic partners. If a student was torn between the popular guy with a six-pack and the awkward geek with an endearing smile, this program seemed to be able to peer into her heart and divine which guy (or gal) she *truly* liked.

I had not set out to develop a dating gizmo for undergrads. After my experience in Geneva studying patients like Huguette, I wanted to test the power of positive emotions on the brain in a systematic way. Huguette had used her love of painting to overcome the brain damage caused by a massive stroke. I saw how the associations she had formed with this beloved vocation literally improved the function and plasticity of her mind. The results of our work together were impressive, but ultimately anecdotal—like any patient's case history.

I wanted to see whether her experience and similar ones I observed with other patients in the neurology ward weren't isolated incidents but rather illuminated some general feature of the brain. I wanted to see whether positive emotional stimuli like love and passion (for a sport, for instance) could improve brain functioning in all people.

Most of the neuroscientists I knew had focused their interest on the opposite side of the emotional spectrum—the Dark Side. Much research had been done, including by some of my colleagues in Geneva, on how negative stimuli sped up the reaction times of certain brain regions. Subconscious priming experiments were conducted where patients were flashed an image of a snake or a spider at a speed that was too fast for people to consciously register but not too fast to escape notice by the amygdala, a brain region that is exquisitely attuned to threats.

Derived from the Greek word meaning "almond"—because of its small, oval shape—the amygdala is buried beneath the cerebral cortex in one of the most ancient parts of the brain, the limbic system, and is designed to receive and act on information about threats blazingly fast, long before such information reaches our consciousness. From an evolutionary standpoint, being alert to a negative stimulus makes perfect sense. If I am an early human foraging in the jungle, I need to be able to rapidly discern whether that long, dark object lying on the forest floor is a twig—or a snake. I need also to be able to see a person in the brush and detect that he is a stranger with hostile intent, so I can run away.

This evolutionary response happens through what the neuroscientist Joseph LeDoux calls a "low road"—a direct emotional pathway designed to elicit defensive responses without conscious thought. It is like an expressway connecting the visual input of the threat from your eyes to the amygdala, triggering the hypothalamus to turn on the "fight-or-flight response," the body's self-defense mode.

All this occurs in the blink of an eye—or about a hundred milliseconds—which is preconscious. (Conscious thought processes kick in at about three hundred milliseconds, or a third of a second.) That's why you might flinch or jump or raise your arm to a threat almost automatically, before you even perceive what it is you're responding to.

The function of the amygdala was perhaps most dramatically illustrated by the neuroscientist Ralph Adolphs's patient S.M., whose amygdala had been destroyed because of a genetic disorder. As a consequence, she could not experience any fear. Incidentally, this was a frightening state of affairs for S.M., since without the ability to detect threats she could not steer clear of dangerous situations, which explains in part why she was the victim of several violent crimes.

Yet as much as the amygdala registers fear, it really is out to detect *saliency*, changes in the environment that are worth noticing. The brain, as a general rule, is wired to detect change. Typically, when a situation is stable it is safe. When it is rapidly changing, not so much. This is why the amygdala is famous as a threat detector when, in fact, it picks up on all kinds of changes, positive or negative.

I once conducted a study with epileptic patients who had electrodes implanted into their amygdalae. They were flashed subliminal messages of both negatively and positively charged emotional words. As predicted, the negative words triggered the amygdala's famed threat detector, but what was fascinating to me was that the positive words also triggered the amygdala—just not as quickly. (And by "not as quickly" I mean a difference of a few hundredths of a second.)

The results suggested that, just as we are hardwired to detect and react to danger, we are also hardwired to respond to the opportunity for positive experiences, things we don't want to move away from but toward. The need for love might be less immediate than the need to avoid danger, but it is by no means a luxury. As we've learned, humans evolved because of love, and we evolved *to* love. So perhaps love had its own ancient pathway—another one of LeDoux's "low roads"—to the brain.

The "Love Machine" was designed to find out if this was true. Here's how it worked: A participant, say the student who visited me that day at Dartmouth for dating advice, would supply the program with the names of the two people she was interested in. Let's say: Blake and Shiloh. Then the test would begin. Her screen would flash. She could see the flash but could not detect that she had just been subliminally primed with the name of Date #1, Blake, which appeared on the screen for twenty-six milliseconds. That is not enough time for the brain to consciously perceive the word, yet it is enough time to relay a subliminal message that activates the amygdala and triggers whatever emotions are associated with the name Blake.

Once this subliminal association is triggered, the participant then completes a series of lexical tasks—sorting out real words from fake ones. By carefully tracking her response times, we can measure tiny differences that statistical analysis revealed to be significant—and meaningful. When the student was primed with Date #1, Blake, she recognized the real words almost 20 percent faster than when primed with Date #2, Shiloh. And randomizing the order—so that Shiloh appeared first—still showed the same speedy reaction to Blake.

But did this necessarily mean that she had an unconscious preference for Blake? What if, instead, she preferred Shiloh and the positive associations triggered by that name actually *distracted* her from the lexical task, making Blake only seem like the favored date? To eliminate this possible confound, I also conducted a "Love Machine" experiment on women who declared that they were madly, deeply in love with their partner. I put the names of the women's partners in the Love Machine along with those of a friend whom they had known for the same amount of time as their beloved. This would ensure that their brain was not simply responding to the familiarity of one name over the other. The result: People performed significantly better on the lexical task after being primed with the name of the person they undoubtedly loved.

Now my question was why. Why was this happening? Why should love improve how quickly a person can read? I guessed that it had to do with the way the brain is interconnected. When the name Blake flashed before the student's eyes and excited neurons in the brain because of the name's positive associations, it activated the brain's "reward" system. As the chemical messenger dopamine poured out from several areas, including the ventral tegmental area and the hypothalamus, it sent a rush of joyous energy coursing through not only the regions that process blissful feelings but also other connected areas, like those that help us parse written language.

The student didn't *decide* for any of this to happen—the reaction and its effects were not something within her will or control or even conscious knowledge. The test, in other words, *revealed* her true feelings, her genuine preference, the fact that her brain had made positive associations with Blake that did not exist for Shiloh. A colleague summarized the upshot of the study this way: "I guess, 'When you know, you know—even when you don't know.'"

The question then became: Why were these feelings inscrutable to her? Why did she need the "Love Machine" to unlock them? This program actually worked similarly to an implicit association test (or IAT), which measures whether people have an unconscious bias favoring one gender or race, for example, over another. Such tests can reveal feelings that are buried deep within—those that you might wish to conceal, even from yourself.

Yet, just like the research my colleagues in Geneva were doing on negative emotions, these tests are typically focused on the Dark Side, unpleasant unconscious reactions, the kinds that are involved in discrimination. Bias is something we need to control and root out. But love is something that we often need to set free. Frequently, it is our unconscious preference ("what the heart wants") that will make us happiest. As Blaise Pascal put it, "the heart has its reasons that the reason does not know."

The problems, the Romeo and Juliet–style drama, usually occur when something gets in the way. Interestingly, when I told the students their results, the response was often some variation of "I knew it!"

"So why did you need the Love Machine?!"

Most students, if they were honest with themselves, had a gut feeling about which guy or girl to go with, but their frontal lobes—which include regions that act like "parents" in the brain, telling us "don't do that"—were standing in the way.

The student sitting in my office would feel a sense of empowerment when she learned that the Love Machine had confirmed her gut feeling. I gently reinforced this feeling and reminded her that the decision to act—or not—on that news flash from her brain was hers, and only hers. She smiled, and walked away from our session with a triumphant attitude—chin up, her two girlfriends following behind.

The Deep End

I knew from our experiments with the Love Machine that when people were primed with the name of their lovers they seemed to think differently, or at least more quickly. This suggested the possibility that love, as an emotion, might be more complex—which is to say, *smarter*—than anyone previously expected. But we didn't yet know exactly what was going on in the brain to account for these differences. We could only speculate about the regions involved or the mechanism behind it. To know more, we needed to peer inside the brain itself.

Functional magnetic resonance imaging (fMRI) is a technique used by neuroscientists to understand the biological basis of different psychological states. Since its introduction in the 1990s, fMRI has played a pivotal role in pinpointing brain areas involved in various cognitive social functions and behaviors. When a brain area is more active, it consumes more oxygen, and to meet this increased demand, blood flow increases to that area. The fMRI essentially charts this process, allowing us to see in high-resolution detail what parts of the brain activate in response to varying stimuli.

In a follow-up experiment, I had thirty-six participants, all women, take the Love Machine test while I scanned their brains using fMRI. This time, I asked them to supply me with the names of someone they had a passionate romantic connection to (as measured by a popular psychological scale), as well as the name of a friend (whom they were not physically or intellectually attracted to), and a hobby they felt passionate about (like tennis or writing, for example).

The results were interesting on a number of levels. First, in terms of the subliminal priming effect, I saw that both groups of people who had been primed with the names of their beloved partners and passionate hobbies enjoyed a significant performance boost on the cognitive lexical task compared to those primed with only friends. And the more they reported feeling in love with their partner, the faster their reaction times.

Next, I looked at what was happening in the brain as the participants took the test. This is where things got *really* interesting. I hypothesized that love would primarily trigger the so-called emotional brain, which comprises the ancient parts of the limbic system as well as the dopamine-hungry "reward" system that psychologists have always associated with love. I saw that all these usual suspects were indeed activated by the love and passion primes.

But these were not the only regions to be intensely activated by love. I also saw that love—as well as passion (for a sport or hobby), but *not* friendship—unexpectedly triggered more sophisticated parts of the brain, like the bilateral fusiform areas and the angular gyrus. These are higher-order brain regions known to be involved in conceptual thinking, metaphorical language, and abstract representations of the self—not exactly regions one readily associates with matters of the heart.

This was astonishing. And one of these regions, the angular gyrus, appeared very recently in our evolutionary history and had evolved along with the traits that make us human: creativity, intuition, autobiographical memory, complex language, experiential learning, imagination, and thinking outside the box. (Part of Einstein's genius, it has been theorized, might have something to do with the fact that he had an unusual angular gyrus.) Why would this region light up so intensely for love?

The angular gyrus hadn't been activated for other positive emotions, like joy and surprise. Might this suggest that love is not merely a feeling but also a way of thinking?

The Map of Love

I was stunned to discover that love and passion triggered these regions that we thought of as being remote from the emotional brain. Was I the only one seeing this data? Or had the evidence been hiding in plain sight all along?

I decided to do a meta-analysis, a big group study of the few previous fMRI studies that had been conducted on love, drawing on results that had been reported in earlier papers and also on supplemental data that researchers hadn't thought to include in their articles—but which might offer clues. The goal was to make a kind of map of love in the brain, to get a full picture of how this complex human phenomenon worked.

My coauthors and I spent weeks in front of the computer, digging into the weeds of the methodology of earlier studies. When we finished crunching the numbers, we found that love seemed to activate twelve specific brain regions. These included not only the usual suspects—the brain's "reward" system and the subcortical regions that control emotion—but also some of the most sophisticated areas I had found in my initial fMRI study, brain regions in the cerebral cortex that are in charge of high-level cognitive functions such as self-representation and body image.

We then compared the brain map of romantic love to that of companionate love (the kind we feel for friends) and the only other type of love that had been studied extensively by neuroscientists, maternal love. All twelve regions in the "love network" were activated by the three different types of love, but the intensity and pattern of activation were different. First, romantic love triggered both the brain's pleasure centers and the cortical regions that manage our sense of self, like the angular gyrus, much more intensely than friendship.

Maternal love was quite similar to companionate love, except it activated the subcortical periaqueductal gray matter (PAG), a brain area that is concentrated with receptors for hormones called oxytocin and vasopressin, which are important in bonding, among other functions. These receptors are also associated with compassion and, interestingly, with pain suppression. This could suggest that, on top of the abundant joy that comes with loving a child, there is something uniquely painful about the experience that requires a natural pain reliever, so that mothers are able to better feel and even absorb the pain of their children. This might ring true to any mother who has night-weaned her baby or sent a kid off to college.

I was fascinated by these results. It was now clear that love played a more complex role in the brain than anyone could have reasonably guessed. But what struck me most about this neural map of love was not its elaborate shape but the fact that it was shared by everyone. People often feel like their own love story is unique, but on a biological level, love looked the same no matter who felt it. Regardless of where you were born, whether you are gay or straight, male, female, transgendered, if a person is significant to you, they will light up this network in the same essential way.

Research in evolutionary and social psychology had suggested that romantic love was a cultural universal, something that existed in every human society. And this study seemed to explain why. Love wasn't some outgrowth of modern civilization, some culturally mediated social construct, but rather an essential, universal feature hardwired into human nature. Now that we had found the map to love, I wondered where it would lead us. Could we use this data to help people find and sustain healthy relationships? What could be learned by tracing love to its deepest roots?

5

Love in the Mirror

Love is the most powerful tool in the universe.

—HOWARD "H" WHITE

For twenty minutes we sat next to each other in complete silence. Then he turned to me and said, "If I start snoring, punch me."

As far as pickup lines go, this one had shock value. I giggled. Then I saw a professor near us who was slumped in a chair.

"He's snoring. Do you want me to punch him, too?"

We both laughed.

"I'm John, by the way."

I arched my eyebrows as if to say: "Duh." We had never met, but Dr. John Cacioppo needed no introduction, at least not at a social neuroscience conference. I had been steeped in his articles since grad school. I had no idea, however, that he was so good-looking.

He had olive skin, with dusky gray hair and a wiry physique. And instead of aging him, his thick mustache had the effect of making his face appear even friendlier, emphasizing a broad smile.

It was a very early morning in January 2011. We were in Shanghai, although we could have been in New York or Milan—the corporate venues where scientific conferences are held always have the knack of making somewhere feel like anywhere.

I had just flown in from Geneva, where I was working as a research professor of psychology for the Swiss National Foundation. In just a few short years since my earliest breakthroughs at Dartmouth, my work had become a lodestar in the expanding universe of social neuroscience. And now Dr. Love was being invited to speak at conferences all over the world.

As we sat there, jet-lagged, chitchatting about research over oolong tea, I had no idea that this conference was going to change my life. And I had very nearly missed it. Twenty-four hours earlier, I was lying in bed in my apartment in Geneva with a temperature of 102. I had been sick with the flu for nearly a week, and the night before my flight I showed no signs of getting better. I wrote to the conference organizer, a psychologist who studied empathy—a curious fact, since empathy was noticeably lacking in his dismissive response to the news of my illness.

"It's too late to cancel," he said. "The programs are already printed."

I had never missed a scheduled talk before, but the fever kept rising. I was too sick to crawl out of bed, much less board a flight. I canceled my ticket, buried my throbbing head under a pillow, and passed out.

In the morning, I awoke, feeling strangely . . . better. My fever had disappeared. I checked the time. I could still make a flight to Zurich, then catch a red-eye to Shanghai. I called Swissair; there was exactly one seat left on the next flight to China. I threw on some clothes, then grabbed my computer, a pair of leather heels, and a black blazer. Then I called the taxi.

"I'll give you a giant tip if you can get me to the airport in twenty minutes," I told the driver.

It took forty-five minutes. I sprinted to the gate and was the last person aboard before they closed the door. I sometimes shiver when I think about what else I would've missed if I hadn't made that flight.

As the academics filtered into the conference hall that morning, John and I quickly slipped into an easy conversation. He knew my work, which meant a lot to me, since it was John who cofounded the field of social neuroscience back in the 1990s. He was particularly interested in research I was doing on the unconscious. We talked about statistical significance and the response to positive stimuli, all the while smiling at each other. Is this, I wondered, how neuroscientists flirt?

The Lone Ranger

John was an academic celebrity. He had written about twenty books, his articles had been cited more than a hundred thousand times, his research had received tens of millions of dollars in grant money. But it was not his résumé that drew me to him but his deep and empathetic hazel eyes. Although he spoke quickly, his words barely able to keep pace with his thoughts, he was also a great listener, and those eyes always made you feel seen, heard, and understood.

John Terrence Cacioppo was born in the small East Texas city of Marshall on June 12, 1951. Flat, dry, and hot, it was another planet from my native Alps. But, like me, John had Italian roots. His grandparents had come over from Sicily around the turn of the century.

His family's hardworking immigrant ethos permeated everything John did. His students would always feel amazed—whether they left work at 8:00 P.M. or came in at 6:00 A.M.—to see that the light was still on in John's office. He used to tell them, "I will never ask you to work harder than I do." Even as John's stature grew, he was not above canceling his dinner plans to help students input survey results if it meant making progress on a study.

A math prodigy as a boy, John was the first member of his family to go to college. At the University of Missouri, he studied economics. But he also had a talent for debate. People told him he would make a great lawyer. For sport, he would sometimes argue one side of an issue, convince you he was right, only to flip sides and successfully argue the opposite position. This exercise, John used to say, "proved to me that I knew nothing." The older he got, the less interested he became in winning arguments and the more interested he became in finding the truth.

Around this time, John happened to meet an experimental psychologist. John asked the man some question about why people do the things they do.

"No idea," the psychologist said. "But that's an empirical question. We can find out." This was John's "aha" moment, when everything shifted; it was a moment not unlike falling in love. He applied to a PhD program in psychology at Ohio State. There he stood out for his intellect and his nonconformist attitude.

That John chose to study *social* psychology and would go on to cofound *social* neuroscience had a lot to do with a brush with death he had as a young man. As the story goes, he was driving at high speed on a one-lane road when a horse darted in front of his car. John swerved and lost control of the steering wheel. In the moments before the crash, what flashed before his eyes were not his many achievements in school but rather the people he loved. And he realized something basic that informed the rest of his research: Social connections were what mattered most in life, what gave life its deepest meaning.

John first made a splash in the late 1970s with his work on persuasion, understanding how people change their minds or make a choice in response to new information. With his best friend at the time, Richard E. Petty, John created the Elaboration Likelihood Model, which will be familiar to anyone who has ever taken Psych 101.

This framework divides our psychological response to persuasive information into two routes: *central* and *peripheral*. The central route involves deliberate thinking, weighing pros and cons, and usually results in the kind of attitude change that is long-lasting. The peripheral route is more emotional, more affected by gut feelings and external factors and biases; this kind of persuasion is less likely to last.

The paradigm has been used to investigate attitude change in everything from advertising and political polling to health care. And the theory bears a strong resemblance to the ideas that one of John's friends, the Nobel Prize–winning psychologist Daniel Kahneman, would popularize much later in his best-selling 2011 book *Thinking, Fast and Slow*.

While John first distinguished himself in psychology, he never wanted the field to be an island. He was interested in mathematics, medicine, technology, physics. And he thought psychology could be a "hub discipline," uniting disparate fields around a shared ambition to understand what it meant to be human. When John coined the term *social neuroscience* with his colleague and dear friend from Ohio State, Gary Berntson, back in 1992, most of his colleagues in psychology thought of the new field's name as an oxymoron. A gulf separated the social and biological perspectives in psychology. John wanted to build a bridge across it. It was a visionary move. Today *neuro-* is hybridized with almost every subject in the academy, and social neuroscience is a thriving field.

John loved Ohio State and, after getting his PhD there, became a pillar of the psychology department. He also loved Buckeye football—so much that it became a distraction. He liked to say that was why, in 1999, he left Columbus for the University of Chicago, which proudly lacked a football team and hired John to lead its social psychology department and create the Center for Cognitive and Social Neuroscience.

It was at the University of Chicago that he began the lifesaving work for which he is perhaps most famous: studying the dangers of loneliness. In paper after paper, John showed that loneliness is a dangerous condition—one that is contagious, heritable, and as deadly as smoking a pack of cigarettes a day. More than anyone on the planet, John knew how important social connections were to everyone's physical and mental health. Yet, sadly for him, this knowledge had not led to a happy love life. He had been twice married and twice divorced. He used to say that he was a good partner on vacation. But as soon as he got home, when normal life resumed, he was reunited with his true love: work. And his human relationships always suffered.

The Glamour

It was a long day in Shanghai: talks, posters, Q and As, mingling, note-taking, swiping tiny bottles of water from banquet tables. I gave my talk about the mind-expanding power of love. John gave his keynote about the mind-numbing danger of loneliness. I thought about congratulating him afterward, but he was mobbed by admirers.

I saw him that night, at a reception at a cocktail lounge, the Glamour, in the former French Concession. The harsh fluorescents of the conference center were now replaced by chill LEDs and lantern light. I walked up a stairway of illuminated steps into an elevated lounge framed by huge windows of the city skyline and the Huangpu River. The place was crowded, noisy, but there were some quiet corners where you could settle in for a nice long chat.

John sat at the bar, schmoozing with the conference hosts. At some point, he got bored and turned to his left, where he found me. He yawned theatrically, recalling the moment we met.

"You know what to do if I fall asleep, right?"

We both laughed. We went on to talk about so much that night. And as we spoke, I felt a tightening, a collapsing of the space between us. I was finishing his sentences. And my French accent, as thick as fondue, which had tripped up so many Americans, was giving him no trouble at all. He understood me so well. We said "me, too!" and "I agree" so often that it became embarrassing. When people have smooth, harmonious conversations like these while they are connected to EEG machines and seated side by side, their brain waves actually synchronize—a phenomenon neuroscientists call brain-to-brain entrainment.

At some point, John asked if I was single.

I blushed. "I'm married to my work," I told him, using a well-practiced line.

"Me, too."

Then John told me about his relationship troubles, his divorces, how much pain he felt he had caused those who loved him, despite his best efforts. He said he thought he might be better off alone.

"Not lonely," he said, "but by myself." John said that his devotion to work meant he couldn't always give a significant other the attention and time they deserved, or needed. And he didn't want to risk hurting someone he cared about. He also didn't want the inevitable conflict that could come when a partner tells you to work less, or that you'll burn out, or that you should call it a night. I related to him. Our histories were different, yet our different paths had led us to this same place of prizing solitude and personal freedom.

Looking back, it's unbelievable to me that neither of us was struck by the irony of our situation, that John and I, which is to say "Dr. Loneliness" and "Dr. Love," were not practicing what we preached. Our research, from opposite ends of the spectrum, emphasized the human need for social connection. And yet both of us had the hubris to think we could go it alone. Were we like physicians who smoked, perhaps thinking, irrationally, that because we understood a disease it couldn't harm us?

As our conversation deepened, the colleagues around us began to fade away, one by one, retiring for the night. We swiveled in stools toward each other, leaned in—all telltale signs of attraction.

If I had stopped to think about what was happening, I could've realized that, neurologically and biologically, I was already falling for him. As we connected, dopamine was flooding my brain's reward circuitry, creating a feeling of euphoria. My heart rate was elevated. Adrenaline was expanding the capillaries in my cheeks, causing me to blush. My norepinephrine levels were spiking, allowing me to focus intensely on our conversation with an excited, nervous energy—the kind that makes time seem to warp.

"Wow, it's already midnight," John said.

We had been talking for three hours straight. I had an early flight in the morning. We gathered the few remaining professors in the lounge and stumbled out of the bar together.

"After you," John said, opening the door for me. As we walked into the dark street, I looked up like I often do. The moon glowed bright, directly above our heads like a beacon. You had to crane your neck ninety degrees just to see it. On all my nights stargazing, I had never seen a moon like that. John snapped a photograph with his iPhone. Then we bid each other good night and retired to our separate hotel rooms.

Smoke and Mirror Neurons

Why did we click? Now that I have some distance I can speculate on the reasons. A lot of it had to do with how much we had in common. Alikeness, it turns out, finding common ground, is a strong predictor of mutual attraction. Studies show that when people play a simple "mirror game"—in which they copy each other's movements—they find the other person more attractive to a surprising degree.

John and I certainly had a lot of things in common: We were both in love with our work; we were both early birds, waking before sunrise to get a jump on the workday; we shared an Italian heritage, an egalitarian spirit, a childlike sense of humor, and so on. But these commonalities didn't entirely explain our connection. I had just met John, yet he felt to me not like a stranger with shared interests but more like a long-lost relative. I felt as though I recognized something of myself in him. We could finish each other's sentences and round out each other's thoughts.

From a biological perspective, it made sense: When we become *significant* to another person, when we share an identity on a deep level, we can harness the power of the brain's *mirror neuron system* (MNS) to anticipate their actions and even their intentions.

As you might remember from chapter 1, neurons are the basic building blocks of the brain. Yet a subset of these neurons—in our motor system and in the regions responsible for language and autobiographical thinking—fire both when you act and when you witness the same action performed by someone else. Just think about how you sometimes get pumped up watching other people play a sport you enjoy, or start to laugh just by seeing someone crack up—even if you haven't heard the joke. The neurons involved in those activities are firing even though you're experiencing these things only vicariously. Most people think that "putting yourself in someone else's shoes" involves a slow process of cognition, but the foundation of that empathic response happens instantaneously on a cellular level in the brain.

Mirror neurons were first discovered by chance in Italy at the University of Parma in the early 1990s. The researchers, led by the world-renowned neurophysiologist Professor Giacomo Rizzolatti, were investigating the premotor cortex, a brain region that helps control the action and intention of our motor system. They had surgically inserted single neuron microelectrodes into the premotor cortex of a macaque and were monitoring the electrical activity—*in vivo*. Every time the monkey moved to grasp an object, its machine would beep.

One day, Rizzolatti's team of researchers were feeding the monkey peanuts (a widely accepted approach in animal studies). All of a sudden, they realized that every time a researcher picked up a peanut to give to the monkey, the neurons in the monkey's brain were activated. The monkey stood motionless, yet the machine beeped as if the monkey were moving to grab the peanut himself. The same thing happened when the researcher brought the peanut to his mouth. The monkey's neurons were *mirroring* the action that he was observing, in this case the action of feeding oneself.

Interestingly, these neurons not only mirrored the actions of others but seemed to "sense" and "understand" the intent and the motivation behind the action. For instance, when the researcher raised the peanut to a place *near* his mouth, the mirror neurons did not respond. Only when he intended to eat the peanut did the mirror neurons get excited and fire in the brain. Later on, Professor Rizzolatti and his team performed a series of rigorous experiments confirming and expanding on this serendipitous finding. Eventually, I had the extraordinary opportunity not only to collaborate with Rizzolatti on research about mirror neurons in human subjects but also to connect the pioneering scientist to an EEG machine and witness his own brain in action.

This was back in 2007, during an early chapter of my career, when I had been asked to direct an EEG lab under the distinguished neurologist and neuroscientist Dr. Scott Grafton. Rizzolatti came to visit our lab, which had recently moved from Dartmouth College to the University of California at Santa Barbara. After adjusting 128 little spongelike electrodes on top of the great scientist's head, we recorded Rizzolatti's electrical brain activity while he looked at pictures on a computer screen of people grasping simple objects (from a cup of coffee to a glass of water) or moving with different intentions (raising their arm to drink or not to drink).

Our results were stunning. For the first time, we discovered that the human mirror neuron system could understand intentions of other people subconsciously—in the blink of an eye. We extended the effects Rizzolatti had found in nonhuman primates to EEG studies in humans. We then replicated our results with a larger pool of human subjects in a series of experiments and published our findings in peer-reviewed journals.

Rizzolatti's discovery of mirror neurons had unleashed a tidal wave of scientific research worldwide. Mirror neurons were said to be involved in everything from language to autism. And I wondered whether they were also part of what allowed couples in love to understand each other so deeply and to anticipate each other's actions.

To answer this question, I turned, perhaps counterintuitively, not to couples in love but to opponents on a tennis court.

The tennis court gave me an experimental setting that was far easier to control than the unpredictable stage on which most human romances play out. But just like a good partner in a relationship, a good tennis player must be able to read his adversary. As fans of the sport know, a serve by a professional like Serena Williams, Naomi Osaka, or Roger Federer can reach upward of 140 miles per hour. At that speed it takes less than four-tenths of a second for the ball to reach an opponent's racket. How do elite players prepare to return a ball that's coming at them so fast, as they do hundreds of times during a match?

Is it just pattern recognition, the result of thousands of hours of court time, that allows them to know that a certain rotation of the hip or flick of the wrist causes the ball to go here versus there? Or is there something else that helps these athletes put themselves in their opponent's shoes at the critical moment? At the 2013 U.S. Open, John McEnroe said, "If there is something you don't want to be on a tennis court, it's predictable." Yet professional players—in order to be remotely competitive—have to predict their opponents' actions thousands of times during a match. What special power allows them to do it?

To find out, I invited experienced tennis players into the fMRI machine and showed them video clips of tennis serves that cut off at the moment a player made contact with the ball. As a tennis player myself, I was not surprised that, compared to beginners, the participants were able to guess where the ball would land with impressive accuracy. What did surprise me was that watching these serves, even just on video, activated the MNS system—as if the participants were serving the ball themselves.

Then I showed the same group of elite players a different set of videos. As with the first set of videos, the second set showed a tennis player serving a ball. But there was an important difference. The server in the video did not know where he was going to direct his shot until after the ball toss. Once the ball was airborne, a research assistant standing nearby shouted instructions to either hit the serve down the middle of the court or aim for the far corner of the service box.

This meant that the server's body language could not have sent any meaningful cues about where he intended the ball to land. He was like the researcher in Professor Rizzolatti's lab standing with a peanut in his hand with no plan to eat it. And, in this case, the MNS stayed dark. What this meant was that mirror neurons fired only when the viewer perceived *intention* in the actions of another.

The MNS seemed to be the subconscious mechanism by which these elite players could predict where the ball would land, but it was a fragile one. If these participants overthought, deliberated, or tried to rationalize their initial reflexive judgment, their guess was no better than a beginner's.

So, for a tennis player to win a match, she had to trust her hardware, her biology, her intuition, the fact that she was *wired* to understand her opponent. I suspected that this was also true of happy couples in love and that relationship problems were often the result of us interfering with our natural ability to read each other's minds and connect.

A New Connection

A few days after the scientific convention in China, I was back in my lab in Geneva. It was a frigid afternoon in January. Snow decked the tops of the plane trees around the lake. The winter holidays had just passed, students were either absent or slowly returning to campus, and my research felt like it was in hibernation. I waited on emails, answers to grant proposals. It was 2011, a new year that felt very much like the old one. I was thirty-six, about to turn thirty-seven. I remember feeling a chill.

I opened a new email draft. I wanted to write to John. But what form of address to use?

"Dear Professor Cacioppo,"

No, too formal.

"Howdy, partner."

Too flirty.

"Remember me?"

Too desperate.

I sighed and simply wrote "Hi John." And then more words came.

> This might sound strange, but do you remember that photograph you took the last night in Shanghai? I've been thinking about that evening. I'd love to get a copy of the picture, if you still have it . . .

Dot, dot, dot: Three ellipses worth a thousand words.

I was starting to realize that our meeting in Shanghai had meant something to me, and I wondered if it meant something to him, too. Maybe he was the kind of guy who connected with lots of different people, maybe the chemistry I felt, which was so rare for me, was commonplace for him, and I was already lost in the crowd of collegial faces.

An hour later, he replied with the photograph and an unrelated little postscript about how he was preparing to give a lecture at a theater in Chicago before a performance of a Sophocles play, and was trying to figure out how to relate it to neuroscience. I knew the play, and I dashed off a few lines for him to use.

He wrote back very quickly: "Not only is she beautiful, but also smart."

OK, now we were flirting. The email chain grew, and soon we were talking at all hours, via phone, via Skype. It was almost as if that conversation in Shanghai never stopped. We talked about the goals we shared in life, how we liked to spend a typical day, the latest research breakthroughs or must-read journal articles.

Since there was literally an ocean between us, the prospect of a second date was complicated. But there was always another academic conference just around the corner. The next one was in Utrecht, in the Netherlands. I flew there for a week. We took a side trip to Amsterdam, went on long walks along the canals, and, on our way to the art museum, our hands touched by accident in the back of a taxi. From that point on we were rarely in each other's company without holding hands.

After the next conference, in Chile, we took a small plane down to Ushuaia in Patagonia, the southernmost city in the world, making jokes about how we were following each other to the end of the earth. We didn't tell our colleagues about our budding romance—for the moment, this was ours to share. We would sneak out of the conference center for romantic dinners. In the morning, we lingered in the airport lounge, dreading the boarding announcement, which always came too soon.

Negotiating a long-distance relationship among two neuroscientists who studied relationships was a curious dance. We understood the intention, the subtext underlying every step we took as a fledgling couple, the effect each action was having on us biologically, psychologically. We knew that when we made eye contact we were activating our mirror neuron system, that when we cuddled we were releasing oxytocin, that when we compared all the ways we were alike, we were in effect measuring self-other overlap, or the degree to which a couple feels as if they are one—an important predictor of the health of a relationship. And yet none of this dampened the sense of excitement we felt in finding each other. Nor did it make us feel especially self-conscious or awkward.

A few months later, I called my mom to tell her about John. She had waited so long for me to find someone, and I had spent so much time pushing that idea out of the realm of possibility. I thought I had chosen a life without love because I wanted to devote myself to science. But as soon as I met John, I realized that I not only had a capacity but a *need* for love—the same need that I had identified in my research subjects but never in myself. As soon as she answered the phone, the truth came pouring out. I realized that I had delayed looking for a relationship because, on some level, I thought I would have to change myself to fit with someone else. "*Maman*," I said, my voice breaking, "I think I have finally found someone who loves me exactly as I am."

6

When the Brain Swipes Right

We loved with a love that was more than love.

—E. A. POE

I fell in love with John's mind. Yet I could not deny that I found him physically attractive: his intelligent eyes, his broad smile, the way he moved, the fact that he was in such great shape. And it makes me wonder, if John were the same person, exactly the same on the inside, but less attractive to me on the outside, would we have clicked the way we did? When I'm in a poetic mood, I answer yes, of course we could have—we could have lived, as E. E. Cummings once put it, "by love alone though the stars walk backwards." But when I am in a scientific frame of mind, I become curious about the precise role that physical attraction plays in forming long-lasting romantic relationships. Is it possible to have the kind of passionate connection that fires up the love network when there's an absence or deficit of physical chemistry in a couple? Can love exist without desire?

Poets, songwriters, and philosophers have posed versions of these questions since the dawn of time, yet clear answers have eluded them. Much of the confusion goes back to how we define love. If you've ever felt intensely and passionately in love with someone whom you find intellectually and physically irresistible, you know you can't easily disentangle your feelings. In contrast, if you've ever had a friend crush, you know that you can "fall for" someone without wanting to sleep with them. You can develop an intellectual infatuation, thinking about a person obsessively, feeling a jolt of excitement when they send you a text. And yet, the idea of physical intimacy does not cross your mind. This describes all close relationships for the small share of the population—approximately one percent, according to recent studies—that is asexual.

Back in the 1960s, the psychologist Dorothy Tennov surveyed five hundred individuals on their romantic preferences. About 53 percent of the women and 79 percent of the men agreed with the statement that they had been attracted to people without feeling "the slightest trace of love"; and a majority of the women (61 percent) and a sizable minority of the men (35 percent) agreed with the statement they could be in love without feeling any physical desire.

To our modern sensibilities, these numbers may seem surprising. Today we barely need to look at the evidence to know that lust can exist without love. But what about the possibility of romantic love without lust? Can true love ever be truly platonic?

That might sound far-fetched, yet when, in 2009, AARP surveyed a nationally representative sample of more than two thousand American adults about their attitudes around love and relationships, they found that 76 percent of respondents eighteen and older agreed with the statement that true love can exist in the absence of "a radiant/active" physical connection. Women, interestingly, were only slightly more likely to agree with this statement compared to men: 80 versus 71 percent. And history provides many interesting case studies showing that this type of connection is possible.

Take, for example, Virginia and Leonard Woolf. They were lovers in every respect, except physically. To Virginia, romantic happiness meant "everything: love, children, adventure, intimacy, work." Leonard could give her most of these things. He was a devoted companion, a friend, a collaborator, a guide, and a source of support during artistic and emotional crises. But he was not a sexual partner; Virginia preferred women. And in a letter dating from their courtship period, she confessed her feelings. "I go from being half in love with you, and wanting you to be with me always, and know everything about me, to the extreme of wildness and aloofness. I sometimes think that if I married you, I could have everything—and then—is it the sexual side of it that comes between us? As I told you brutally the other day, I feel no physical attraction in you."

They married anyway, and for three decades Leonard supported his wife in every way he could. When Virginia took her own life, at age fifty-nine, she left him a note in which she wrote, "You have given me the greatest possible happiness. . . . I don't think two people could have been happier than we have been." What is this, if not romantic love? And yet . . . who could also deny that Woolf was missing something that, for most couples, is a necessary ingredient to a lasting, fulfilling relationship?

This gets us back to the sticky issue of definitions. If you define romantic love in a broad and polymorphous way as just a deep affection and attachment, it is of course possible to love a person without desiring them physically. But, if you define love based on its unique neurobiological blueprint, it is clear that desire is not an incidental feature of a loving relationship but an essential ingredient. This desire, as we will discover, doesn't necessarily *need* to be sexual but it must be physical. By that I mean it must involve not just the mind but the body as well.

Making Love

When you combine desire and love, you go from having a physical experience to *making love.* We think of the former as more about the body, more individualistic, more about fulfilling one's biological desires and needs, more about the now than the future. We think of the latter as less about the body than about the mind or the heart and soul, less about the individual and more about the relationship, less about *me* than *we*. When a couple make love they are intentionally fusing together, communicating mentally and physically that which they cannot find the words for, sharing, realigning and resolving differences, embodying the harmony and fluidity and connectivity that couples so often seek.

Yet on a neurobiological level, the more you look at the dividing line between love and desire, the blurrier it gets. Think of a person whom you find extremely physically attractive. As much as you might believe your feelings are merely physical, with every (real or imagined) touch and kiss, your brain is complicating matters. The pleasure you're experiencing results from the same neurochemicals, from dopamine to oxytocin, that flood your body when you're in love. This is one of the reasons why people may grow attached to those they once considered just a "friend with benefits."

Physical intimacy helps us not just form an emotional connection with our partner. It also makes us feel the importance of the physical body, makes us understand what the literary scholar Joseph Campbell called "the rapture of being alive"—which he believed, more than some vague sense of "meaning," is what most of us are actually searching for in life. The goal, he said, was for "life experiences on the purely physical plane [to] have resonances with our own innermost being and reality."

We experience and react to desire even before we're conscious of what's going on. Let's say you're going for a walk in the park on a sunny day and holding hands with your partner. Suddenly, a beautiful runner crosses your path, and your partner's eyes are drawn like a magnet to the runner's body. In many cases, your partner won't even notice that they are looking until you point it out, usually with an annoyed glance.

"What?!" your partner asks, uncomprehendingly.

We rarely realize the extent to which our gaze, our attention, is automatically and unconsciously directed by the nature of our interest in someone. Using eye-tracking studies, which can identify exactly where a participant is looking, my research team and I have found that when both men and women are shown a photograph of someone whom they find physically attractive, their gaze instinctively falls on that person's torso (even when dressed). But when they look at someone they later say they could imagine falling in love with, their gaze falls directly on the face. And the stronger the potential connection, the more likely they are to focus on the eyes. We knew from previous research that eye contact is one of the most reliable markers of love between couples, but this study showed that people fixate visually more on a person's face (relative to their body) when they are thinking about feeling love.

Maybe this is a component of what people call "love at first sight"? The fact that our eyes are drawn to someone's face, the way that I was drawn to John's when I met him in Shanghai, signals to us that this person may be someone special. The importance of eye contact in loving relationships was indirectly reinforced in 2020 when a team of researchers from the Yale School of Medicine showed that real-time direct eye-to-eye contact awakens activity in a core brain area of the love network—the angular gyrus. In this study, thirty healthy adults (fifteen pairs) were seated across a table from each other. Each partner was asked to gaze at their partner for a total of ninety seconds (alternating every fifteen seconds between direct eye gaze and rest). Overall, these results suggest that reciprocal eye-gaze between partners increases activity in neural circuits that play a key role in love.

Untangling Love and Lust

In 1969, two brave female social psychologists, Ellen Berscheid at the University of Wisconsin and Elaine Hatfield at the University of Hawaii, paved the way to study the mysteries of love and lust with a landmark book, *Interpersonal Attraction*. In the decades that followed, they conducted rigorous descriptive and experimental research that laid the foundation for much of what we now know about the psychology of interpersonal relationships and its cultural evolution around the world. In Hatfield and Berscheid's view, love and lust can be experienced separately or in concert. Physical attraction, they explain, is more often closely associated with the experience of *being in love* with a partner than *loving* a partner.

Another important authority on the science of lust and love is the Rutgers biological anthropologist and self-help author Helen Fisher. Her rigorous work from survey data and fMRI studies led to a groundbreaking theory that love and lust can be seen as distinct phenomena that recruit different brain systems.

In the first brain system, she speaks of the "sex drive"—that involves primitive brain regions, and motivates people to crave and seek out a range of partners. The second brain system in her theory focuses our mating energy on a single person at a time and results in a state of infatuation that feels to the body and brain very much like the experience of taking an addictive drug. And the third system, attachment, creates a "deep sense of calm and commitment toward another person" that enables people to "tolerate that person" over the long haul—it's akin to the kind of bond one develops with a best friend, a companion, or a teammate.

When John and I happened to run into Helen Fisher a few years ago in a hallway on our way to a scientific meeting, we were delighted to finally meet in person, given our shared research interests in the science of love. Though we have different backgrounds and training, she and I agreed that these brain systems can often operate in very complicated ways, especially when it comes to passionate love.

A growing line of neuroscientific research suggests that passionate love and lust might rely on a single interdependent and unified brain network. And this network capitalizes not just on basic mating drives or cravings, which all primates have, but also on complex cognitive energy from areas of the brain that are uniquely human. A deeper look inside the brain reveals that passionate love and lust, while often seen as opposing or rival forces, may actually work together. They exist on a continuum, and understanding how and when they relate to each other can help us become better partners.

Together they may deepen our sense of connection to our significant other. When I scanned the brains of twenty-nine young women who had scored high on a passionate love scale and then asked them a series of questions, I found that the more women self-reported feeling emotionally close to their partner, the more they said they were satisfied with their physical relationship with that partner. These women also showed the greatest activation in a brain region called the insula. I wondered why this was the case.

The insula was once known by the evocative name the "Island of Reil"—after Johann Christian Reil, the German anatomist who discovered it two hundred years ago. Located deep inside the cerebral cortex in a fissure where the frontal lobe cleaves from the temporal lobe, the insula plays a key role in self-awareness and responds to a host of different stimuli: pain, addiction, music, the pleasure we take in food. It is this part of the brain that helps us understand what we crave, whether it's a sandwich or a milkshake or a massage.

Neuroscientists have theorized that the reason why the insula is activated by such a variety of experiences is that one of its important functions (in addition to immune system regulation and homeostasis) is to help us make sense of—and assign value to—bodily experiences. Yet when I reviewed all the existing literature and conducted an in-depth statistical analysis, drawing on twenty studies and 429 participants, I found that desire wasn't lighting up the whole insula but rather a specific isolated part toward the back of this brain region—the posterior insula. Feelings of love, meanwhile, triggered the anterior insula, at the front. The plot was thickening!

This back-to-front pattern matched a broader tendency in the way our brains are organized: Posterior regions tend to be involved in current, concrete sensations, feelings, and responses; and anterior regions are more involved in relatively abstract thoughts or introspective awareness—what we *think* about what we *feel*.

Other imaging studies revealed another region that had a similar, seesaw response to love and desire: the striatum, a subcortical area that's involved in processing rewarding experiences. The ventral, or underside, of the striatum—which responds to inherently pleasurable activities such as sensual touch and food—was intensely activated by desire. But the dorsal part of the striatum—which assesses the *expected reward value* of such experiences—was more activated by love.

In scientific paper after scientific paper, we were discovering that love and desire were activating complementary parts of the same brain regions, reinforcing the idea that they are not necessarily opposing forces but have the potential to grow out of the other. Love was essentially the abstract representation of the rewarding, visceral sensations that characterize desire. It was as if desire were crushed grapes and love the elixir made from them with time and care.

Two to Tango

One method neuroscientists use to test their theories is to look for people who have damage—because of illness, a stroke, or an accident—in the specific regions of the brain they are investigating. These neurological case studies, often based on a single person, offer a rare opportunity to address causation—rather than correlation—between a brain function and a behavior.

I've relied on case studies to make breakthroughs in my research since the beginning of my career. And in 2013, to better understand the insula's role in love, I tapped my connections in the world of neurology to find a patient who had a brain lesion in the insula. I got in touch with Dr. Facundo Manes and Dr. Blas Couto from the Institute of Cognitive Neurology in Buenos Aires, who thought they had a patient who could help me unlock the mystery.

Let's call the patient RX. He was a forty-eight-year-old heterosexual man who, excepting a recent minor stroke, was otherwise healthy. The stroke had left a lesion confined to the anterior insula.

If our prior meta-analysis of the fMRI studies on love was correct, damage to the anterior insula should, in theory, have hampered RX's ability to form and sustain loving connections in some way.

RX had come to his neurologist with typical stroke symptoms, but luckily most of them—headache, facial palsy, speech difficulties—were only temporary. By the time our team met him, he was neither depressed nor terribly anxious. His general intelligence was unaffected by the stroke, and moreover, his social intelligence seemed also entirely preserved. Tests showed that he could easily recognize basic emotions and feel empathy for another person's pain.

We designed a special experiment to determine whether the stroke had caused some subconscious deficit in RX's ability to love. We also recruited a control group of seven healthy men who were similar to RX in age and other demographic characteristics. We then showed RX and the members of the control group a series of photographs of well-dressed women, as one would see in a dating app profile. For each photograph, the subjects made a choice—did they merely find the woman attractive or could they also imagine falling in love with her?

RX looked no different from the control group when it came to the number of women he could imagine falling in love with. But, interestingly, it was taking RX far longer than the control group to make these decisions. When it came to desiring a potential partner, however, RX was just as fast as the control group.

The patient hadn't noticed any deficit in his ability to love, but perhaps his wife had? Because sometime after his stroke, RX and his wife split up. The reasons were complicated, as they often are in a breakup, yet I wondered whether the damaged insula, a pillar of the brain's love network, had played some role in the dissolution of his marriage.

Lessons from the Lab

As neuroscientists continue to tease apart the roots of love and desire in the brain, you might wonder how this line of research can be applied to your own love life. In most countries, the vast majorities of people (78 to 99 percent) describe "a faithful marriage to one partner" as their ideal romantic arrangement. Unfortunately, most long-term couples will encounter some problems with physical intimacy over the course of their relationship. As couples age, they can easily lose the passion that first attracted them to each other. Studies show that physical intimacy sometimes declines as couples get older—markedly so if children are in the picture. Dysfunction in this area is a surprisingly common problem. Forty-three percent of American women and thirty-one percent of men have experienced some kind of difficulty with physical intimacy during their marriage. A third of all Americans suffer from a deficiency, if not a total absence, of physical desire. These issues have been reported to greatly increase the risk of a breakup.

Despite the pervasiveness of these problems in long-term relationships, the bar that couples set for themselves is often precariously high. The majority of couples rate physical intimacy as being "very important" to their relationship, and a healthy love life is, according to the sociologists Sinikka Elliott and Debra Umbersen, seen as a cultural signifier of "marital bliss." What this means is that most people in relationships view love without desire as being not quite complete. And neuroscientific findings seem to reinforce this idea, since the insula is a critical part of the love network and it seems to need a deep emotional and cognitive connection, in addition to a passionate physical connection, to be fully fired up.

Yet I wonder if you could compensate for a lack of physical

chemistry by finding other, nonsexual ways to activate the posterior insula? I'm not talking about some cutting-edge technique like deep brain stimulation that is, at least for the moment, best confined to clinical settings. Instead, a simple behavioral brain hack might do the trick. Remember that the posterior insula is also activated by food. This part of the brain manages the awareness, perception, recognition, and memory of taste. So, if you're having trouble connecting with your partner, why not try your luck in the kitchen? Try creating new recipes together, or cooking together, or sharing a delicious meal together—even if it's just a meatless burger and salad—and focus on the novelty of the flavors, the sensuality of that shared experience, and let your brain, including the insula, do the magic.

Beyond taste, the posterior insula is like a radar for a range of sensory and bodily experiences. Cuddling, hugging your partner, or smelling pleasant fragrances together will also likely activate this region. And, as highlighted by the Swedish neuroscientist India Morrison, the posterior insula's location in the brain circuitry of stress buffering suggests that such pleasant sensory experiences will have a calming and grounding effect on you and your partner.

Walking together, running together, or dancing together are just examples of other activities that will switch on this part of the insula. Studies have shown that couples who dance together are happy together. Not only will you again trigger some of the bodily sensations of pleasure that spark in the insula but you can also reduce stress and increase relationship satisfaction. The lesson to take from this chapter is that your connection to your beloved is a profound one that benefits from cognitive and bodily feedback. Luckily, when it comes to the insula, there's more than one way to awaken your senses.

7

We'll Always Have Paris

How often is happiness destroyed by preparation, foolish preparation!

—JANE AUSTEN

When I woke up on the morning of September 28, 2011, I had no idea that this would be my wedding day. John had been invited to a MacArthur Foundation event in Paris, an international mind-meld of twelve academics from different fields who all had something to contribute to the study of aging. The group included some of the world's top public health experts and psychologists. John came to discuss, among other things, how the elderly could protect themselves against the dangers of loneliness. But he and I were now several months into our long-distance love story and getting good at juggling business and pleasure.

I had taken the high-speed train from Geneva to Paris the previous afternoon. I tried to work on my laptop as the train sped through the Jura Mountains and the vineyards of Burgundy, but I was too distracted by my anticipation. It was crushing to spend weeks away from him, but that bit about absence making the heart grow fonder? That's not just a saying—it's science.

In an interesting 2013 study, people in long-distance relationships were shown to have more meaningful interactions with each other—even though they could only communicate via text, phone, and video chat—compared to couples who saw each other every day. This, paradoxically, resulted in a deeper connection for the long-distance couples. Across social species, we see the power of distance to refresh relationships—even in elephants, who give each other more elaborate greetings after a prolonged absence.

Part of the reason is the social brain's innate preference for novelty. Distance keeps us from taking a person for granted. Distance reminds us what we miss the most about our significant other. No wonder that parting is such sweet sorrow—and reuniting feels so good it hurts.

John had spent that day arguing with his colleagues over the finer points of methodology and the value of this or that intervention. But as the sun set over the City of Love, he excused himself from the group and met up with me for a long walk along the Seine. As soon as we saw each other, we started walking faster and jumped into each other's arms.

Though we didn't have a reservation, we talked our way into a table at the romantic Tour d'Argent, the centuries-old restaurant perched above the river. We sat at a small table with a view of the back of Notre-Dame, all lacy flying buttresses bathed in amber light. The buttoned-up waiter in his three-piece suit looked shocked when John passed up the typical tasting menu, which would include countless courses and interruptions to our conversation. Instead, he ordered a pressed duck to share and two glasses of Champagne.

"Tonight isn't about the chef," he said. "It's about us."

The next day I was working on a research article in the hotel room. I remember the day felt special already because September 28 was my grandmother Mémé's birthday, and I was thinking wistfully about her and wishing she were still alive to meet John, who was now officially my fiancé. Mémé would have liked the way that he had asked me, getting down on one knee. And she would have especially liked that after I said yes, he insisted on calling my father and "asking for my hand." My parents cried tears of joy. They knew that John would cherish me just the way my father and my mother had always cherished each other.

We were so happy to be engaged that we had forgotten all about the small issue of planning the wedding. We thought we'd eventually get around to it. But for the moment, there was too much work to be done.

That day in Paris, John had left the hotel early to meet up with the other conference attendees. After a long morning session, he was taking a break with his good friend, the Stanford psychologist Laura Carstensen. They had last seen each other a month earlier, at Chicago O'Hare Airport, when John was headed to yet another European conference (and another date with me). But a bad snowstorm had delayed his flight.

"Look on the bright side," Laura told him. "If the flight is canceled, you don't have to give your talk."

John suddenly looked very serious. "No, I'm meeting someone over there, someone important."

He started to tell Laura about me, how we had fallen for each other after a chance encounter in Shanghai, how he *needed* to be on that flight. Laura knew about all the wreckage from John's two divorces. It struck her that John spoke about me as though

I was his last chance at love. He said he wanted to get this one right.

Now, in Paris, Laura wanted an update: "How are things with Stephanie?"

A sheepish grin. "We're getting married."

"Wow, John, that was fast! Congratulations! I'm so happy for you. When is the big day?"

He looked momentarily confused.

"Oh, I don't know if we'll do a classic wedding. Steph and I are both so busy. We'll probably just go down to City Hall in Chicago during our lunch break."

"You know, I could marry you right now."

Laura had just gotten ordained as an internet minister to officiate the wedding of one of her grad students. She meant her suggestion purely as a joke—"there was not a shred of seriousness in it," she later told me—but John didn't seem to realize this.

"You mean, like, today? You know, Steph's here in Paris with me. . . . That could actually work!"

He grabbed his phone and began typing a message to send to me. DO YOU WANT TO GET MARRIED AFTER WORK TODAY?

"John, wait—what are you doing?!" Laura said. "Are you *crazy*? These things must be planned out." John didn't seem to be listening. The conference was about to resume. Laura just shook her head and muttered under her breath, "John, you don't know women."

He smiled devilishly. "*You* don't know Steph."

So, *did I want to get married today*? I did a double take when that message flashed on my phone. But it only took me two seconds to reply. "Sure."

Then I dashed out of the hotel room to find a white dress.

Turning Your Back on the Frontal Lobe

By surprising me, John had instantly found a way of making our wedding special and unique, however it turned out. Such a spontaneous ceremony wouldn't work for everyone, of course, but it's worth considering the role that unpredictable events might play in love, and whether we can benefit by making more room for improvisation in our relationships.

So much of our social experience, especially when it comes to romance, has to do with expectations. Maybe we have an image of the person we will marry long before we've met them. Usually, we call this a *type*, an ideal. Or maybe we have in our mind's eye the perfect first date: a walk by the lake, a hike in the woods, a romantic restaurant. When it comes to the wedding—an opportunity not only to proclaim our love but also to show off our good taste and social network—we probably know how it should look. Perhaps more importantly, we know how it should *not* look.

If these expectations set us on a course toward genuine happiness, then they are all well and good. But I would argue that in many cases such plans can become a kind of mind trap, forcing us to pursue a preconceived kind of happiness that we may never reach or that, once reached, may not actually make us happy.

To take just one study proving this point, the Yale psychologist Robb B. Rutledge and his colleagues conducted an experiment in which they had participants set expectations before playing a decision-making game with small financial rewards. Their results show that the ultimate amount of money participants won did not determine their happiness. Rather, what predicted their level of happiness was the difference between their initial expectations and the outcome. If they had no expectation of winning and ended up with even a tiny amount of money, they were happy.

Applying this expectation formula to love relationships, the more we love without expecting any rewards in return, the more we will increase our chance of happiness. This is in line with a large body of research showing that setting realistic expectations leads to greater relationship satisfaction. But adjusting your expectations does not necessarily mean lowering them. It's more about letting go of the social pressure that often drives us to pursue unrealistic expectations without understanding what it is we really want or need, and what we can do without.

The important thing about letting go of your expectations is that it must feel like an act of generosity or faith in your relationship—and not a *sacrifice*. Otherwise, you will likely feel that you're giving up something important to you for the sake of your partner, leading to feelings of resentment or spite, which can spell serious trouble. An interesting study from the Netherlands shows that while couples greatly appreciate it when their partner makes sacrifices for them, when they begin to *expect* such sacrifices they feel much less gratitude and no longer see their partner's sacrifices in quite the same positive light. This finding is in line with a theory I've long held: *Expectations kill gratitude*.

We have a natural tendency to expect help, support, and sacrifice from our partner to some extent, but the beauty of our evolved brain is that we are neurologically wired to control this tendency and, if so moved, to expect less from our partner and give more in return—*intentionally*. The next time you have romantic dinner plans with your partner but they feel "brain dead" after, say, an entire day of back-to-back Zoom meetings, ask yourself what's more important: sticking to the script you had for how that evening *should* play out, or letting it go for both your partner's sake—and your own. You might discover that simply sharing an apple pie or stargazing together from the comfort of your backyard under a warm blanket was actually more romantic than whatever elaborate evening you had planned.

Expectations don't only complicate life within a relationship, they also can get in the way of connecting with someone in the first place. What if you find the "One" but you don't recognize that person as such because you had a different idea of what the person you marry *should* look like? Conversely, what if you stay in an unhealthy or abusive relationship for too long because you and your partner *should* be perfect together, because on paper your partner is the kind of person who *should* check all your boxes? Sometimes the search for that perfect partner can make someone go to extraordinary lengths. Such was the case with Linda Wolfe, a woman from Indiana who married twenty-three times—earning her a place in the *Guinness Book of World Records*—but who never found Mr. Right. Shortly before her death in 2009, she said she was still holding out hope that husband Number Twenty-Four would come along to fulfill her dreams.

These dreams, hopes, expectations, and scripts we write for our future that play out over and over in our minds are managed (in part) by the prefrontal cortex (PFC). Considered for decades a mysterious brain area, the PFC is now seen by many neuroscientists as one of the parts of the brain that make us most human. Located smack in the front of the cerebral cortex, it contains some of our newest hardware, evolutionarily speaking. Its large size and extensive connections with a diverse pool of neighbors enable it to contribute to a broad range of mental functions, including decision making, language, working memory, attention, rule learning, planning, and the regulation of emotions, to name just a few.

Earlier I described this region as the brain's "parents," telling you what you should and shouldn't do. The processing that occurs in the PFC is the closest thing we have to Freud's notion of a "superego."

This part of the brain helps us to distinguish right from wrong, to control and suppress urges, to see silver linings in dark situations, and to make hard decisions and delay gratification, if it benefits us in the long term or benefits some cause greater than ourselves.

It is a little-known fact that the PFC in humans doesn't fully mature until about age twenty-five. This explains why you might have done things at eighteen—a keg stand? a tongue piercing?—that you would never have done a decade later. Yet as much as we need the PFC in order to be functional and responsible adults, there are times when we want to reconnect with our younger selves, not to second-guess our actions so much, to think more about the now, to be in the moment. I think often about the French poet Charles Baudelaire's definition of genius: It's nothing more than "childhood recaptured at will."

I love this idea. And when I am truly living in the moment, this is how I see the world, with the joyful curiosity and sense of wonder that children possess. This is not, however, an invitation to impulsive behavior. We don't want to completely turn off our PFC. That would be a disaster.

If we did that, we would be ruled mostly by our impulses. We would lose a grip on and lack the ability to regulate our emotions, to manage psychological pain, and to complete tasks we had scheduled for the future. We would end up looking at the world less like children, who are eager to learn and discover new things, and more like patients with a lesion, a tumor, or a disease affecting the front part of the PFC (known as the orbitofrontal region). Such patients exhibit a lack of impulse control, have difficulties in meeting personal and professional responsibilities, and make all manner of social faux pas.

One famous case study that illustrates a personality change after a frontal lesion is that of Phineas Gage, a twenty-five-year-old rail worker from New Hampshire. In 1848, he suffered a gruesome injury when an explosion launched an iron rod straight through the left side of his head. Miraculously, Gage survived, but he was never the same. His left orbitofrontal region had been destroyed. And with it went much of his common decency, as the physician who treated Gage, Dr. John Martyn Harlow, later reported. He went from being hardworking, well mannered, "very energetic and persistent in executing all his plans" to "pertinaciously obstinate," "irreverent," and erratic. This transformation was so intense that his friends said he was in fact "no longer Gage."

More recent case studies have revealed how essential the PFC is to our social nature. The French neurologist Dr. François Lhermitte once had a patient in the 1980s with a large tumor in her frontal lobe. She would follow him around aimlessly, grab objects that he was holding for no apparent reason, and imitate his actions without the slightest sense that she was doing anything odd. Similarly, the American neurologist and neuroscientist Dr. Bob Knight once asked patients with damage in their PFC to read the following story to see if they could identify a faux pas.

> Jeanette bought her friend Anne a crystal bowl for a wedding gift. Anne had a big wedding and there were a lot of presents to keep track of. About a year later, Jeanette was over one night at Anne's for dinner. Jeanette dropped a wine bottle by accident on the crystal bowl, and the bowl shattered.
>
> "I'm really sorry, I've broken the bowl," said Jeanette.
>
> "Don't worry," said Anne, "I never liked it anyway. Someone gave it to me for my wedding."

For most people the faux pas would come through loud and clear. But Dr. Knight's patients with damage in this area were completely unable to perceive why Anne's remark could be seen as tactless.

While the PFC is essential to our social nature, sometimes we let it control us too much. Studies show that rumination, recurrent negative thinking, self-focused thinking, and even obsessive-compulsive disorders are associated with changes in the PFC. The prefrontal cortex on overdrive is not a pretty sight. We think less flexibly, we obsess over minute details, make ourselves sick with worry, turn something over and over endlessly in our head. In such cases, we are trying too hard to stay on script, to anticipate what will happen, to plan and perfect.

This kind of PFC-driven overthinking can at times be a roadblock to creativity. When neuroscientists zap parts of the PFC to reduce its influence (using a noninvasive technique called transcranial magnetic stimulation, or TMS), we find that people show a cognitive enhancement as they are better at solving problems or brainteasers, and tend to think more outside the box. With mental training, we can achieve similar results whether that means coming up with a solution to creative problems at work or finding a way of turning a scientific conference into an impromptu wedding party. Not only are we more creative when the PFC is kept in check, we're also in a better mood. Studies show that the less we ruminate, the higher our subjective life satisfaction and happiness.

So the question then becomes: How do we keep the PFC in balance? How do we benefit from all its useful aspects—the way it helps us plan, save, and control unhealthy or harmful tendencies—without letting it rule our life, causing overthinking, anxiety, and other problems? In other words, how do we go off-script? How do we say no to FOMO, the fear of missing out, and say yes to JOMO, the *joy* of missing out?

Popular antidotes to combat negative overthinking include

breathing, meditation, and mindfulness exercises. Such techniques have in recent years gained in scientific credibility. Two of the people who helped change attitudes toward meditation and mindfulness are the American neuroscientist Richard Davidson, a pioneer in the art of reining in the PFC, and the French Buddhist monk Matthieu Ricard, a master in finding wonders in solitude and strength in self-introspection. Together they have formed a friendship and collaboration with the Dalai Lama. They have conducted rigorous experiments in Davidson's laboratory at the University of Wisconsin–Madison (and in other laboratories around the world) in which they used EEG and fMRI to study the brains of Tibetan monks and other students of meditation. Their results show how highly trained practitioners (with over nine thousand hours of lifetime practice on average) can control negative thought processes, accept every feeling as it comes in a nonjudgmental way, and modulate the activation of various brain areas, including the PFC—the region that, as Davidson described it, is "absolutely key in emotion regulation, as the PFC is a convergence zone for thoughts and feelings."

Yet it's not only eminent monks with years of training in meditation who can benefit from such techniques. In 1999, Davidson and his colleagues contacted the CEO of a biotech company and suggested that they teach his employees mindfulness meditation and then assess how it might affect some measures of physical and mental health. Forty-eight employees volunteered to take part in nonjudgmental moment-to-moment awareness training, which is also known as mindfulness-based stress reduction (MBSR). Every week for two months, the employees participated in a two-and-a-half-hour session. The researchers measured the volunteers' brain waves (focusing on the PFC) before and after the training period. At the end of the eight weeks, the results were clear. Compared to a control group, MBSR volunteers showed a 12 percent drop in their anxiety symptoms and a shift from right to left PFC activation. Interestingly, the PFC in the right hemisphere of the brain tends to process negative emotions, whereas the left PFC specializes in positive emotions—an indication that the mindfulness training was working.

There are now numerous mindfulness apps and behavioral therapies that can help ruminators become meditators and gain new insights not only into their physical and mental health but also into how their brains function. In time, they can learn to suppress the impulse to obsessively focus on negative past and negative future events, putting the PFC in a more peaceful place. What devotees of meditation and mindfulness techniques find is that being in the moment can quite literally clear the mind of unwanted and unnecessary negative thoughts and replace them with more positive, constructive thinking.

Finally, nature has also been shown to have a potent effect on reducing rumination and regulating PFC activity. For instance, a 2015 study performed at Stanford showed that participants who went on a ninety-minute walk through a natural environment showed reduced neural activity in the parts of the PFC that foster rumination. By immersing yourself in nature or making intentional changes to your mindset, you might open yourself up to experiences that could be richer, deeper, and more meaningful than anything you could have ever planned or imagined.

Hard Truths

Since that first night in Shanghai, John and I had been falling deeper and deeper in love—but soon rumination started to kick in. There was one big biological fact separating us: our ages. John was sixty. I was thirty-seven. And he worried whether marrying a much younger woman was a good idea. I don't think he cared about what it *looked* like, the assumptions some people would make based on their own scripts for a "proper relationship." I don't think he was thinking about himself at all. Rather, he was worried about me.

If something happened to him, I would be alone, and, much worse from John's perspective, I would be susceptible to loneliness. John *knew*, perhaps better than anyone on the planet, what loneliness could do to a healthy brain. And he was sensitive to the actuarial reality of our circumstances. If we tied the knot, odds are he would not be around to celebrate my sixtieth birthday.

"Since when did you become the prince of darkness?" I asked him.

He smiled, but then his eyes became serious. We were in a small restaurant in some random German city whose name I've already forgotten. Dating across an ocean didn't always feel like glamorous jet-setting. Sometimes it was exhausting. This was a few months before Paris. John had taken two flights and a train to see me, and we only had one evening together. It started out as a beautiful night, getting lost in a foreign city, free-flowing conversation, romantic and unplanned, spontaneous. But after we settled in to dinner, the conversation took a turn from the now, from this moment, to the looming future.

"We may be in love, but we haven't committed yet," John said. "We're still on the outside of this. I know, because I've studied it, the kind of loneliness that overwhelms a widowed spouse. The idea that you could live with that for decades—I can't in good conscience put you in this position."

I'm stubborn. I told him to forget it. I told him that I could never let something as seemingly trivial as age dictate our love. Besides, he was in great shape. He was so happy with me. I was so happy with him. We *fit*. And we had gotten so enmeshed in each other so quickly that even the idea of putting some outside constraint on our love felt to me like an artificial intrusion. Looking back, I can't help but think that it wasn't only our age but also our academic specialties that influenced our positions. I was arguing for the joyful power of love, and he for the destructive power of loneliness.

We decided to take a week away from each other. No phones, no Skype. Some emotional distance. I went exploring caves with a girlfriend in southern France. And he went on vacation to Peru. We both spent our trips going through the motions of having fun, trying to distract ourselves from the fact that we missed each other so much it was hard to breathe. At the end of the week he sent me a photograph of his left hand. He had put a silver band on his ring finger. "I'm yours," he wrote.

Wedding Crashers

"Could one get married in a blazer?" I asked myself while canvassing all the boutiques of the Left Bank in search of what felt like the only white dress in Paris. Meanwhile, the whole scientific conference was metamorphosing into a wedding party. We couldn't rent a spot on such short notice, so we decided to do an impromptu ceremony in a random corner of the Luxembourg Gardens, near our hotel. John asked Dr. Jack Rowe, a professor of public health at Columbia and ex-CEO of the insurance giant Aetna, to give me away.

"But I haven't even met the bride!" Jack said.

Laura Carstensen would officiate. The University of Pennsylvania sociologist Frank Furstenberg, armed with his iPad, served as our photographer. The hotel chef whipped up a wedding cake with just a few hours' notice. An economist from the group couldn't help calculating how much a wedding in Paris would have cost us if we had planned it. "Do you know how much money you're saving!?" Honestly, the thought hadn't even crossed our minds.

As John and I stood there, united in our love, I found myself looking at the people around us, many of whom I had only just met. They were beaming. None had expected to end up attending a wedding, but each played a role in the moment, each felt a sense of taking part in something special.

As Laura finished her speech and we got ready to exchange vows, I heard a French voice suddenly shout *"Attention!"* Two policewomen had come by to inform us that the gathering was in violation of several park rules. It was absolutely forbidden to be on the grass. As John and I stood arm in arm, and I held my bouquet of flowers, feeling a bit awkward, the francophones in the wedding party pleaded with the officers to let us finish the ceremony.

"Have a heart!" shouted a passing tourist. "This is Paris!"

The policewomen discussed the situation among themselves for what seemed like a long time. After deliberating, they decided to let us quickly finish the ceremony, provided we dispersed right after we exchanged vows. But we immediately had to step off the grass. So, like some choreographed dance routine, the entire wedding party rotated themselves in perfect synchrony, stepping over the elegant metal border separating the grass from the gravel path.

And then we exchanged vows.

About two weeks later, we officially got married in downtown Chicago, during our lunch break (just as John had imagined). Both our spontaneous wedding ceremony in Paris and our official wedding in Chicago remain two of the best moments of my life, but I have to confess, in my deepest heart, I feel like our spontaneous ceremony marked the first day when John and I became man and wife.

8

Better Together

Together we can do so much.

—HELEN KELLER

Ruben Toledo was from Havana, Isabel from a village in the Sierra Maestra, but they met in high school in New Jersey, two Cuban immigrant kids who dreamed of making art together. He fell in love the first time he laid eyes on her silky black hair and alabaster skin, but it took her nearly a decade to see this debonair goofball, with his pencil mustache and head full of big ideas, as something more than a friend. In the meantime, they explored New York City, explored their creativity, pushed each other in new directions. He could draw, and she could sew, and they both could dance. Disco was the thing in the late 1970s, and they spent nights hustling between Studio 54 and Andy Warhol's Factory, making friends who would soon sell their clothes at chic boutiques like Fiorucci and Patricia Field.

As their fashion business grew, so did their mystique. With Ruben's help, Isabel created dresses that floated like kites and slightly off-kilter workwear that made a generation of creative New York women feel like they had found a second skin. Isabel's crowning moment came in 2009, when Michelle Obama chose one of her dresses—a glittering golden sheath with a matching coat made out of wool lace—to wear to her husband's inauguration.

An artist friend said that Isabel was the fabric and Ruben the needle—they couldn't make such beautiful things if they were not stitched so tightly together. Ruben's interest in his wife seemed bottomless. He once thought he had drawn her portrait ten thousand times. They often hinted that much of their communication was not verbal. She would drape fabric on a mannequin, then write a few words down on paper, and sketches would magically appear in the margins. It was mysterious, this connection—they couldn't really explain it, and often they reached for metaphors. She called him her faucet—ideas would just pour out. She was the sieve, picking, choosing, coming up with the concept, the feeling. She dedicated her autobiography to him: "FOREVER RUBEN who is the warmest part of me I have no control over."

When Isabel Toledo died of breast cancer in 2019 at age fifty-nine, her memorial service became a celebration of the love she shared with Ruben. In a room of their weeping friends, he read a farewell note to his wife. "You know how well we fit together—like two misfit puzzle pieces that magically unite forever. . . . I see everything and everyone through your unforgettable eyes." Their love seemed no longer to belong only to them; it was a shared thing, a monument of affection that touched everyone who knew them.

The Lover's Edge

The Toledos are just one of many examples of how love can make two people feel like they are more than just the sum of their parts. Another striking case is the scientific power couple Marie and Pierre Curie. They met in a lab at the Sorbonne in Paris when they were poor, young chemistry students, yet by dint of their unbreakable covalent bond, they built a rich life together, in which they both felt hypnotized, as Pierre put it, by their shared "scientific dream."

That dream took many forms—from investigating radioactivity to discovering the elements radium and polonium, a breakthrough for which they won a Nobel Prize in 1903. Each of the Curies understood they couldn't have achieved so much without the other. After Pierre died in 1906 in a tragic carriage accident, Marie felt that the only way forward was to continue their passion. She went on to win a second Nobel Prize, in 1911, and never remarried.

What is extraordinary to me is that you can have all the intellectual benefits of a symbiotic relationship like the Toledos or the Curies even if you don't share the same line of work or passion in life with your partner. In fact, even couples with vastly different work lives report that being with their partner makes them think faster, makes them more creative, makes them somehow a better *version* of themselves. My research shows that this is more than a feeling, that in many measurable ways couples in love enjoy cognitive benefits that are lost on those who don't have such a passionate connection.

We already knew from the Love Machine experiments that merely thinking about your beloved (even on a subconscious level) could improve how quickly you can read. And many other studies have suggested that love is good for the mind in unexpected ways. Researchers have found evidence suggesting that love facilitates creativity and the type of brainstorming or motivational synergy that leads to innovation; that the so-called love hormone oxytocin enhances creative performance; and that love priming (for instance, asking participants to imagine taking a long walk with their beloved) helps them tackle intellectual challenges that are not related to their relationship. Prior research also revealed that the more people report being in love, the more creative they consider themselves.

These were neat studies, and had neat effects, but I wondered if we could learn more about the fundamental nature of love or its evolution—how it worked, why it brought people together. What I was interested in was whether love could boost skills that specifically helped us navigate the social world, an ability that psychologists call *social cognition*. This would help us understand whether love had a function beyond just perpetuating the species and consolidating parental support.

To begin to answer this question, I conducted a series of experiments comparing how well people in love could anticipate the actions of their partners compared to strangers. My earlier research on the mirror neuron system had shown that tennis players could use their connection with their opponent to anticipate where a tennis serve would fall. Now I wanted to see whether people in love could use their connection to predict each other's behavior.

As expected, I found that people can read the intentions of their significant others much better than those of strangers. Not only that, but the more in love you are, the faster and more accurate your predictions will be. This ability could explain in part why couples like Ruben and Isabel or my parents—or John and me for that matter—seemed able to communicate without even speaking.

Yet I wondered if it was really love or just familiarity—the simple fact of knowing someone very well—that accounted for this ability. After all, when you're in a loving relationship, you've probably watched your partner make the same faces and do the same actions thousands of times. Might this store of experiences explain the lover's edge? To find out, I created a sense of familiarity in strangers by showing research subjects repeated actions from the same stranger dozens of times. Yet this exposure had no effect on their ability to predict what they were about to do—suggesting that it was indeed love, not familiarity, that accounted for the edge.

The question then became whether the cognitive benefits of love would generalize to other social relationships. Could being in love not only help us interact with our partner but also improve how well we understand the emotions and intentions of other people, a skill that psychologists call *mentalizing*?

In a fascinating experiment, Robin Dunbar and Rafael Wlodarski from Oxford University found that making research subjects think about their beloveds made them significantly better at assessing the mental states of strangers, compared with subjects who were made to think about a close friend.

Interestingly, while earlier research had shown that women are innately more empathetic and better at reading the emotional states of other people, this study showed that when people are primed to think of their beloved partner, men significantly outperformed women when it came to assessing *negative* emotions in particular. In an evolutionary context, this increased sensitivity to negative emotions might have helped our ancestors protect their partners and detect external threats to the relationship.

Right Angles

You'll notice that couples in love often refer to the other person as their soul mate or "better half." They talk about themselves using the pronoun "we" instead of "I." They stand very close together, often linking their arms and hands automatically, as if becoming a single unit were the most natural thing in the world. For couples deeply and passionately in love, the usual, transactional give-and-take that characterizes other social relationships doesn't apply. They experience their partner's victories as their own; they *feel* their partner's pain in defeat or loss. They think nothing of giving up something valuable, or enduring some discomfort, if there is a net benefit to the relationship, even if it accrues only to their partner.

This is more than empathy. This is the result of what psychologists term *self-expansion*. A theory developed by the husband-and-wife social psychologist team Arthur and Elaine Aron, self-expansion assumes two interlocking truths about human nature: (1) people have an innate drive to broaden themselves—by following their curiosity, honing their abilities, or exploiting new opportunities, and (2) the primary way they do this is through close relationships, particularly romantic relationships, in which the idea of the self ("me") expands to *include* the other person ("we").

Self-expansion gives you the ability to experience someone else's identity as your own. As the distinguished psychologist Barbara Fredrickson has written, when couples experience this aspect of love, "The boundaries between you and not-you—what lies beyond your skin—relax and become more permeable." She describes it as "a transcendence that makes you feel part of something far larger than yourself." Albert Einstein experienced this with his first wife, the Serbian mathematician Mileva Marić. His sense of self was so enmeshed in hers that whenever they were forced to separate, he seemed not quite himself. "When I'm not with you I feel as if I'm not whole," he wrote to her. "When I sit, I want to walk; when I walk, I'm looking forward to going home; when I'm amusing myself, I want to study; when I study, I can't sit still and concentrate; and when I go to sleep, I'm not satisfied with how I spent the day."

The self-expansion impulse explains why, in experiments, people are particularly attracted to potential partners who have the qualities that they themselves would like to possess, qualities that are part of some "idealized self." This type of thinking—in which one thing (the relationship) comes to represent another (the self)—is essentially metaphorical. And metaphor is the specialty of the cognitive region of the love network called the angular gyrus.

As you might remember, the angular gyrus was one of the few higher-order cortical regions in the love network—meaning it is found far above the more primitive subcortical regions of the emotional brain. It is a small triangle located just behind your ear in the parietal lobe. The area is of tremendous interest not just to me but also to lots of other neuroscientists, although its function is still somewhat mysterious. It is intriguing that it responds so strongly to love, since it does not seem to respond to other positive emotions, like joy or surprise.

Patients with damage to the angular gyrus often lose the ability to process words or do basic arithmetic. When we stimulate this area, meanwhile, some patients, including those I tested when I was in Switzerland, reported out-of-body experiences. We know that the angular gyrus is found only in great apes and humans, which means that it was one of the more recent parts of the brain to evolve. We also know that it is triggered when people think creatively, make unexpected associations, or connect the dots in new ways.

In my own research, I've discovered that the more a person in love perceives their self and their partner's self to be overlapping, the more activation we see in the angular gyrus. What is potentially exciting is that the angular gyrus also helps us manage not only metaphors but also other aspects of language (in addition to spatial attention, numbers, and autobiographical data such as our self-image). So the fact that it lit up like a Christmas tree when I scanned the brains of people in love could explain why people in love perform faster (compared to a control group who report being "out of" passionate love or being in a friendly relationship with their partner) on reading tests or tasks measuring creative ability or mentalizing skills. Was this the love network's secret power source?

Coming Home

I probably didn't need to do an fMRI analysis or any other scientific research to understand the concept of self-expansion or the facilitating effects of love on the mind. I could just look to my own relationship. What John and I experienced was nothing short of a cognitive metamorphosis—a profound expansion of our concept of who we were that began to include the other person, just as the Arons had predicted.

Because I had never been in love before I met John, and never even had a serious boyfriend, I wondered how strange it would feel to move in together. But the strangest thing about it was how natural it felt. Though I had spent thirty-seven years alone and was suddenly sharing my bed with someone every night, I didn't need an adjustment period. I didn't feel as though I had entered somebody else's life. I felt like I had come home.

Ahead of my arrival, John had meticulously prepared his closets and emptied a couple of drawers for me—thinking that would be plenty of room. In reality, he had underestimated the number of shoes I had in my suitcases. I made the most of the space and donated all the clothes that didn't fit in the space. I didn't need anything more to be happy now. I had him.

Plenty of people in a relationship desire some "healthy" distance from each other, an independent work persona and home persona. They would get tired of each other if they spent every waking hour together. But we felt that we had spent too long without each other to waste even a minute more apart. Our appetite for being together seemed insatiable: We ran together, we did the laundry together, we went food shopping together, we brushed our teeth together.

And, of course, we worked together. I joined John on the faculty of the University of Chicago, where I became the director of the Brain Dynamics Laboratory and an assistant professor of behavioral neuroscience in the department of psychiatry at the university's Pritzker School of Medicine. John and I settled into an incredibly close existence, cowriting articles, mentoring the same cohort of grad students, sharing an office (THE CACIOPPOS was written on the door), even sharing a desk. We got a dog, a Chinese shar-pei named Bacio ("kiss" in Italian), who liked to cuddle under our feet as we worked.

Both of us had been extraordinarily productive as independent scientists. Yet we found that by working together, we could make new connections and come up with better ideas faster. We felt more motivated than ever before, more open to new collaborations and new research paradigms. Yet while we thrived in this intensely close relationship, we noticed that colleagues often didn't get it. I felt a chill sometimes from other professors, who looked askance at me because of my age difference with John, or the fact that I wanted to share an office with him, or that I had chosen to take his name.

Though I had over fifty publications under my maiden name,

Ortigue, I loved John's last name, Cacioppo, which reminded me of the Italian side of my family, to which I've always felt close. Plus, call me old-fashioned, I thought it was a romantic gesture to take his name. For me, it had nothing to do with gender politics—if John had been a woman and we had a same-sex relationship I feel like I would've still wanted to take my spouse's name. People told me it would damage my career, set a bad example for other female academics. I couldn't wrap my head around that idea. Why should the name a person chooses have an impact on how their work is perceived? This was when I became aware of something painful, something that people who had more experience with romance know well: Though there are only two people in a relationship, their opinions are not the only ones that matter.

After I watched some of my ideas for novel experiments be torn to shreds by my new colleagues during lab meetings, John and I decided to run a secret experiment. At the next meeting, he would present one of my study ideas as his own. I sat stunned as I watched the same colleagues who had earlier been so dismissive of my work now praise it to high heaven.

"I'm glad you liked that," John said. "But you should actually tell Steph, since she's the one who came up with it."

At a certain point we decided to stop caring what other people thought. We were not going to allow them to write the script of our love story, label us, or force us to relegate our relationship into our private lives. So many people have ideas about the proper place of love. Think about the old cliché about "fools in love," which suggests that somehow people in a passionate relationship always have their heads in the clouds and are thinking only of themselves. As study after study shows, that's not remotely true. Love sharpens our mind, improves our social intelligence, and makes us more creative together than we could ever hope to have been alone.

9

In Sickness and in Health

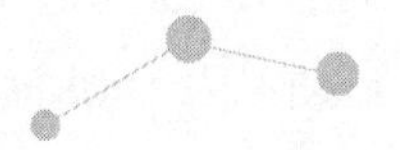

Have enough courage to trust love one more time and always one more time.

—MAYA ANGELOU

I remember there was so much sunlight streaming into our home office that it felt like we were on vacation, like it was the French Riviera outside our window instead of Lake Michigan. The year was 2015. We had moved from a cute and cozy house in the leafy historic district of Chicago, where John had lived for many years, to a spacious new apartment across from Lincoln Park. The building was like something out of a dream, with a gold-columned entrance and a swish lobby designed by the French architect Lucien Lagrange. It reminded me a bit of Paris, and it felt more like a hotel than a condo, with a staff of kind and caring doormen and security guards who made us feel secure and coddled. We had bought the place almost on a whim, after John surprised me one idle Sunday by suggesting we take a look at an open house. Yet it felt in keeping with our spontaneous, unscripted new life together that we would fall in love with it, and just up and move.

John had gotten past the point in life where he cared about how other people might judge him, or whether he "looked the part" of an academic. He had spent most of his life living for others; now he wanted to be his authentic self. He had his eye on a muscular yet graceful two-seater sports car. We took it for a test-drive; it was the automotive equivalent of love at first sight.

People who didn't know him just assumed he chose that car to show off, or to confound expectations, or because he was experiencing a midlife crisis—but John loved that car simply, loved its beauty and power. It spoke to his true self. I rarely saw him happier, saw him smile wider, than when we were all alone driving down the highway and he would step on the gas (while still respecting the speed limit) and let that spectacular engine roar.

Plus, he knew how much *I* liked it.

It had been four years since our wedding—four years of being happily in love, passionately in love, productively in love. We still worked as hard as ever—science, after all, was the basis of our relationship, the fascination we shared that sparked everything. But we also knew how to have fun. Our research made us popular on the speaking circuit. CEOs flew us out to corporate events on private jets; our calendars filled with John's book tours and award ceremonies. We attended scientific events at the White House and the National Institutes of Health. We consulted about love or loneliness for Fortune 500 Companies, NASA, the CDC, the U.S. military.

We had been so busy from the earliest days of our marriage that we never found the time to go on a honeymoon, so we decided that we would celebrate in a small and simple way every single day—

toasting with our morning cup of coffee, watching a cooking show or sports, going for a run by the lake, or playing tennis together. Whenever we were making travel or dinner reservations, if someone asked whether we were celebrating a special occasion, we never failed to answer: "Our honeymoon!"

"Congratulations!" the waiter or flight attendant would say. "When did you get married!?"

"Four years ago."

Everyone just laughed. It may sound sappy, but I honestly think most people who met us found our love story inspiring. We had become living proof of our science, and we somehow managed to make the expected unexpected by cultivating mystery and introducing surprising moments into our daily life. John often surprised me by leaving little notes on my keyboard—some variation of *"I love you."* I would wake up early to surprise him in return, leaving a Post-it on the mirror in the bathroom. *"I love you more."*

John's heartbreaks, my lifelong solitude, the way it all tied into our research about the dangers of isolation and the necessity of social connection—it made our relationship into something that felt larger than us. I think that, for some people, when they thought of true love, it was us they were thinking of.

That sunny day in our apartment had begun like any other. We were sitting at the "his-and-hers" workstation nooked into the corner of our home office, which we had custom-built so we could spend every day side by side. A framed poster on the wall read PARIS IS ALWAYS A GOOD IDEA, reminding us of our improvised wedding. Bacio was curled up under John's feet, as always.

We had nothing special planned for the day. I expected that after putting in our usual twelve or thirteen hours of work, we would exercise, then have a drink on the balcony, then John would cook dinner, and we would clean up together. Maybe afterward we would watch the sunset, sitting on two chunky leather armchairs in front of a wall of windows, as the airplanes gathered over the lake, waiting for the signal to land from Chicago O'Hare. At moments like these, I often thought about a quote from that great aviator turned writer Antoine de Saint-Exupéry: "Love does not consist of gazing at each other, but in looking outward together in the same direction."

This wasn't everyone's idea of a dream life, but it was ours. And then, in a snap of a second, it fell apart. John's cell phone rang. I remember thinking that it was strange because several minutes passed without him saying a word. He just listened. And then he turned to me with tears in his eyes and said, "I'm sorry."

He had thought it was a toothache, this strange, persistent pain in his cheek. He had an unusually high pain threshold, almost never complained, so I was worried when he kept talking about how much it was bothering him. After a couple of weeks, we went to his doctor, who thought it was nothing serious, and told him to see a dentist. Yet the dentist could not explain it either, and the pain persisted. Finally, we went to an ENT, who decided to do a CT scan. "Just a precaution," the doctor said. He'd call us if anything was wrong. And then we got the call.

Our first response was very emotional: We cried, we held each other. But then, within the first hour, our training as scientists kicked in. We did a literature review, we learned all about the rare type of cancer John had (stage IV salivary gland), the odds of his being alive one year after diagnosis (horrifyingly small), the newly developed therapies that offered slivers of hope. The oncologist at our local hospital told us to find a specialist.

"I really have no idea what to do with this," she said.

Her honesty was admirable, if not exactly comforting. She sent us home with a binder full of material on how to cope with stress and find social support. John gave me a funny look.

"Oh, great," he said. "Psychology homework."

"Maybe they cite our research?"

We both cracked up. At least we could laugh in such moments!

We eventually found an amazing team of doctors at the University of Chicago Medical Center. A world-renowned oncologist, Dr. Everett Vokes, would be treating John, and a decorated U.S. Army surgeon, Dr. Elizabeth Blair, would perform the operation. This pair had made international headlines for saving the Chicago chef Grant Achatz's life—and his sense of taste—after he was diagnosed with a virulent form of tongue cancer. When we heard that they were on our side, we felt a flicker of confidence, despite the bleak diagnosis, a sense that if we worked hard, if we put our heads and hearts together and tapped every scientific connection we had, we could beat this.

Dr. Blair covered her mouth with her hand when she saw the CT scan of John's face, how large his tumors were, how the cancer had already spread to several lymph nodes. She did not sugarcoat his chances, or the risk of the surgery she was proposing. She knew she was talking to fellow scientists. "It is what it is." She said she would try her best.

We went to the beach the day before the surgery, to take some "before" photos of John's face. We didn't know what the surgery would do to him. Would he lose an eye? Would his face be paralyzed? John was trying to think of everything. We expected the best but planned for the worst, adapting a coping mechanism called "defensive pessimism," proposed by the social psychologist Nancy Cantor.

Just before we left for the hospital, I saw him go around the apartment and futz with the timer on each one of the dozen electric votive candles we had. He was setting them to turn on the next day, so that if, God forbid, he didn't survive the surgery, I would come home and see the light.

The operation ended up taking eight hours. Dr. Blair looked exhausted when she came into the waiting room to tell me that it went well. She was able to flip open John's cheek and cut the cancer out from the salivary gland while threading her scalpel around the nerves and muscles of his face, preserving his vision and features. She drew, with a few confident swipes of her pen, the outlines of the procedure. I had never seen something so terrifyingly beautiful.

They sent John home with a drainage tube near his incision to prevent fluids from building up, and a monocle to protect his eye. People probably thought he looked like an android out of science fiction, but John didn't care how he looked. He decided he'd spend the first day home from the hospital shooting pool with his brother in the lounge of our building, and just smiled whenever somebody stared.

As little as he cared about his appearance, he was disquieted by the possibility that he wouldn't regain the full function of his facial muscles, which were so critical to his ability to convey emotions. John had done work with electromyography, a technique that records the activity of muscles, which he used to analyze facial expressions in his experiments. He knew the facial nerves like the back of his hand. And he would sit in bed every night as he recovered, practicing to regain his nerve function—blinking, furrowing, smiling.

"I think it's coming back," he would tell me.

Within a few months, John had regained all function in his facial muscles and the surgery had barely left a trace, just a slight asymmetry that you could hardly notice. We even took an "after" portrait to go with our beach photographs.

As with Grant Achatz, Dr. Vokes tried to up John's chances of survival by putting him through what he called the "Trilogy." First the surgery, then a doubleheader of chemotherapy and radiation for seven straight weeks. That meant we had to live in the hospital. I dressed us in matching gowns and cozied up the hospital room with personal photos, flameless candles, a small side table, and cushions from home. I made a habit of wiping down our private space every day and spraying a disinfectant—my way of protecting John in his immunocompromised state. The nurses loved what we did to the room, and they particularly loved the smell.

"Is that Chanel?" one asked after hearing my French accent.

"No," I said, laughing. "It's Lysol."

The university hospital was only one block away from my classroom, so I could walk from the hospital to teach and then come back to be with John around the clock. The nurses knew that "visiting hours" didn't apply to me. We were that inseparable. I slept at first on the chair next to John's bed, but eventually they gave me my own mattress, attached to John's. Soon I moved into his hospital bed. When the night nurse would come by at 4:00 A.M., she often asked us, "Who's the patient?" It was a joke, but it wasn't that far from the truth. This was happening to both of us.

When John was able, he sat in on my class while receiving chemo, with his pole and a nurse nearby. He was too weak to participate—but the sound of neuroscience echoing through the lecture hall brought a smile to his face and his presence inspired my students. When he could, he made his way to the lectern himself.

Once, right in the middle of the Trilogy, John insisted on giving a lecture that he had scheduled before the diagnosis. None of our colleagues had really laid eyes on him since his treatment began, and now they were stunned to see his gaunt face and skeletal body. But he made a joke, got through it. Dr. Vokes came, too. I thought it was because he was curious about neuroscience, but we later learned he was there because he was afraid John wouldn't make it through his talk. He was in such pain, yet you wouldn't have known it looking at him.

Afterward, I asked John why he bothered. "Why put yourself through all this? Why not conserve your strength?" I thought he was doing it for our friends and our students—he wanted them to see that he was still OK, he wanted them to be inspired by his resilience, inspired that such a thing was even possible.

He looked at me, somewhat confused.

"No, that's not it," he said. "I did it for you."

I buried my head in his chest and wept. *I did it for you.*

In that moment, I began to understand one of the differences between the passion we might have for our work and the love we feel for a significant other. These things might share a neural basis, they might light up similar regions of the brain. But had I been just a solitary, unattached neuroscientist in love with her work who was facing John's brutal diagnosis, facing all that pain and suffering, I don't think I would have had the strength. For whom would I be fighting? That might sound harsh, but it's also realistic. Yet for John, I felt like I would have done anything, gone to the ends of the earth, endured as much pain as a person could handle, even given my life. This fact made me realize the significant difference between love versus the passion we have for our work or some other pursuit that's a core part of our identity. While the latter might define some people or give them a sense of purpose, it may not be sufficient to keep them fighting in the face of such struggle.

I realized, before it was too late, that one's identity cannot rest solely on work; that we need to love others to survive.

Love: It Does a Body Good

There were several moments during John's brutal treatment when I wondered whether our love played a role in protecting him. This was not some kind of hocus-pocus wishful thinking. A growing body of research from social neuroscience and other fields shows that love literally makes us stronger—not just emotionally and cognitively, as we've already discovered, but also physically.

Let me count the ways. Compared to single people, those in satisfying, healthy long-term romantic relationships sleep better. They have better immune function. They exhibit fewer addictive behaviors. They suffer fewer recurrent strokes. They even have a better survival rate for some diseases (including some cancers).

Some of these improved survival rates can be attributed to the simple fact that you have someone else looking after you. Research has shown that people in a relationship might help their partners detect skin cancer earlier, for instance, by drawing their attention to a suspicious mole.

Yet having another pair of watchful eyes cannot explain why people in relationships have better survival rates after high-risk surgeries. A 2012 study examined 225 adults who had undergone a coronary artery bypass grafting. Remarkably, the researchers discovered that married patients were 2.5 times more likely to be alive fifteen years after the surgery than single patients who had undergone the same procedure. And it wasn't just the daily physical presence of another person. Those who rated their marriages "highly satisfying" had an even higher survival rate—3.2 times that of single people.

Lab experiments help us begin to understand the "why" behind these impressive statistics. When researchers monitor couples' vital signs while they discuss problems in their relationship, they can see clear evidence that the greater the quality and satisfaction of the relationship, the better the vital signs. And the more our cardiovascular system reacts to stressors like marital arguments, the more susceptible we are to *atherosclerosis*, which is the underlying biological process involved in many types of heart disease. This recent insight is in line with a large body of research on the health benefits of romantic love. For instance, in the late 1970s, a study of ten thousand men showed that if they felt loved and supported by their partner, they had a lower risk of chest pain (or angina), even in the presence of high-risk factors.

Love has not only a calming effect on stress but it also promotes healing. When Janice Kiecolt-Glaser, a clinical psychologist and professor of medicine at Ohio State, gave couples tiny blister wounds on their arms and then asked them to either talk to each other in a loving supportive way or to rehash a recent conflict, she and her colleagues discovered that the wounds of couples who acted more kindly and lovingly toward each other healed 60 percent faster than their more combative counterparts. In another study, they also found that couples who reported more positive interactions—and whose wounds healed faster—had higher natural levels of oxytocin in their blood. This discovery highlights the powerful connection between oxytocin and the immune system, which indirectly supports our collective intuition about the healing power of love. Preventing or reducing inflammation is important in light of recent evidence that systemic inflammation could be associated with increased risk of cancer and cardiovascular disease.

Not only is there evidence that people in healthy relationships reduce their risk of harmful diseases and promote healing, they also literally feel less pain when their significant other is touching them or is even just in the same room. That's why I knew I could never leave John's side during this treatment, and I was so grateful I had the kind of job and support system that allowed me to make such a decision. To be with your partner when they are suffering is more than just some psychic balm—it actually changes the biological reality of whatever medical experience they are going through. When the love network is turned on, it activates the brain's reward centers, releasing oxytocin and a cascade of other hormones, neurochemicals, and natural opioids that help our body heal and our mind deal with pain.

One of the most powerful ways of activating the analgesic power of love is through physical touch. When the University of Virginia neuroscientist James Coan gave small electric shocks to participants in healthy committed relationships, he discovered that those who held their partner's hands experienced significantly less pain—not only in terms of their self-reported *perception* of that pain but, according to fMRI scans, they also had less neural activity in parts of the brain that register threats, like the hypothalamus.

What was fascinating was that in *troubled* relationships this protective effect entirely disappeared; the unhappily coupled women experienced just as much pain holding the hand of their partners as if they were completely alone. Such results are consistent with theories about how supportive social relationships reduce the stress response in the body's autonomic nervous system, which regulates the healthy function of our internal organs.

All this research not only shows the healing potential of love but the importance of *quality and satisfaction* in all our relationships. The fact that you check the "married" box when filling out forms, or that you share a bed with someone every night, means in itself absolutely nothing to the brain and the body—or at least not much. Rather it is the *nature* of the connection between you and your partner that determines whether you will reap the health benefits of love.

An "SOS" from the Brain

When it comes to our physical well-being, the true power of love may not be what it causes but what it prevents. One of the most important things that love can do, it turns out, is shield us from the ravages of chronic loneliness, that state of social deprivation whose dangers to our mind and body John discovered in his pioneering research.

The first thing to know about loneliness is that, while it hurts, it's very much designed to help us—a bit like an alarm system. Evolution has sculpted the human brain to respond to biological mechanisms called *aversive signals*. Some of these we experience every day. Hunger, for instance, is triggered by low blood sugar and motivates us to eat. Thirst helps us find water before we become dehydrated. Pain is also an aversive signal: It helps us avoid tissue damage and encourages us to take good care of the physical body.

John figured out that loneliness also works as part of this biological alarm system. But it alerts us to threats and damage to our *social* body. Loneliness increases our motivation to bond with others—it's the brain's way of telling you: You're in social danger, you're on the periphery of the group, you need protection, inclusion, support, and love.

Although the pain of loneliness is ultimately meant to help us seek out or repair meaningful social connections, ironically, its immediate effect is to make us hypervigilant to social threats. John and I called this "the paradox of loneliness."

Think about one of our early human ancestors, cut off from the tribe, alone and frightened, wandering in the jungle. *Whom can she trust? How can she find the way back into the fold? Where, exactly, does the danger lurk?* Such a state of heightened social vigilance is extremely helpful if you're trying to find your way home in a hostile environment. But what if you're just alone in your apartment, staring at your phone, feeling this way night after night after night? This is when loneliness turns from a lifesaver into a life wrecker.

In social species ranging from fruit flies to human beings, social isolation decreases life expectancy. Scientists used to believe that it was the risky and harmful health behaviors that loners engaged in that accounted for their poorer health outcomes. But evidence is building that loneliness *itself*—and not the things lonely people do—has damaging effects on our health through the way it changes our brain chemistry and sets off a cascade of biological booby traps.

A meta-analysis of seventy studies involving more than three million participants who were followed for an average of seven years showed that being lonely increased the odds of an early death by 25 to 30 percent—roughly the same level as obesity. But unlike obesity, there is very little understanding about how loneliness kills and what we can do to protect ourselves.

When we look at the characteristics of lonely people, we find that there is actually very little that distinguishes them from the nonlonely. Outwardly, they look no different—they weigh no more or less on average than the nonlonely; they are just as tall, just as attractive, just as well educated. Perhaps most surprisingly, they even spend roughly as much time with other people, and are just as adept navigating the social world. In other words, there is absolutely nothing wrong with lonely people except the way they feel about their relationships, or lack thereof.

John always emphasized that loneliness is a *subjective* measure of isolation. You can be married and lonely. You can be at a party and feel lonely—even with a hundred of your closest friends. (And, yes, social media is also a kind of party where you can easily feel lonely.) Conversely, you can be single and have no friends, you can be a monk floating alone in outer space, and *not* feel lonely. Loneliness usually occurs when people are dissatisfied with their relationship and their *expectations* for social connection fall short of their perceived reality. You crave fellowship, companionship, someone there who understands, who *gets* you, and they're nowhere to be found.

Welcome to Planet Lonely. It's a dangerous place, yet it's home to a huge and growing population. At any time, roughly 20 percent of the United States population—sixty million people—report feeling so lonely that it is a major source of unhappiness in their lives. And these people are not just in psychic pain but also at physical risk. Chronic loneliness accelerates the aging process. It floods the body with stress hormones, causing shorter and less restful sleep. It takes a toll on heart health, increases the risk of stroke, and has been linked to Alzheimer's. Loneliness actually has the power to alter DNA transcription in the cells of your immune system, potentially making vaccines given to lonely people less effective.

Many of these health risks occur because the body has been shifted into a survival mode that favors short-term preservation over our long-term health. Try to think of chronic loneliness as a malfunction in the social brain's alarm system. To be lonely is to have your alarm ringing 24/7. Even when your family and friends come to your door, that annoying alarm is still reverberating in your head, trying to alert you to nonexistent danger, as if the people who want to help you might actually be trying to harm you. In this state, when a good friend asks you a simple question, like "Are you OK?" you might think he or she is implying that you're not.

When the alarm goes off, it doesn't only make you more vigilant about possible social risks but it also triggers security measures. Your brain's primary threat detector, the amygdala, is fired up. It directs power to the brain's security cameras—regions that control our vision and attention.

With the security system in overdrive, even a whiff of smoke means FIRE! So the brain needlessly activates its sprinklers—making tonic adjustments to your neuroendocrine system and triggering the body's fight-or-flight response. Blood vessels expand, energy floods to myeloid tissue in the bone marrow, levels of the stress hormone cortisol spike. All this inflammatory activity is designed to help you, but it's actually what hurts you over time—what causes lonely people to have higher blood pressure, lower viral immunity, worse sleep habits, more depressive tendencies, and more susceptibility to impulsive behavior.

The tricky thing about loneliness is that, to some extent, it's self-reinforcing and even self-fulfilling. The more you think you are lonely, the more you are. Your mind can be the loneliest place on earth, *if* you let it be. In my experiments, I've shown that lonely people are much faster at picking up on negatively charged social words (like *hostile* and *unwanted*) than nonlonely people.

The point is, once we're hyperalert to social risk, we begin to see it *everywhere*. Have you ever learned a new word only to see it pop up unexpectedly later that day in a book or a conversation? That's not a coincidence—the word is appearing just as frequently as before you learned it. Only now you're paying attention. The same process fuels the cycle of loneliness. If you're lonely and inclined to see your friends as foes, you will start noticing more cues that prove your point. When we're lonely, we become prickly, dejected, more focused on ourselves—not exactly the frame of mind that's most suitable to making friends or finding a romantic partner.

If we feel lonely and someone looks at us funny or says something awkward, we are more inclined to take offense, to feel rejected, and to "protect ourselves" from that rejection by dismissing the person as a potential mate. Our decade-long longitudinal study of older adults found that loneliness increased levels of self-centeredness, even after we controlled for baseline levels of self-centeredness in the sample.

Lonely people are not only more likely to focus on themselves but they also have a less firm grip on social reality. They are more likely to anthropomorphize animals and inanimate objects—they ascribe human qualities to pets and tend to see faces in clouds. Think of the film *Castaway*, where Tom Hanks's character, stranded on a desert island, made a companion out of a volleyball he found in the wreckage. He named it Wilson.

Why does this kind of thing happen? When we are lonely, starved of social connection, our parietal lobe is imbalanced, which causes overstimulation in the brain areas it is connected to—including areas that store and interpret faces and bodies. The result is that we become so thirsty for human connection that our minds literally create social mirages.

The Lonely Hearts Club

While I held John's hand through chemo and slept next to him in the narrow hospital bed, I thought a lot about loneliness—and how much love protected us from this silent killer. John was already dealing with a devastating health challenge, an invisible enemy—cancer. But what if he had been dealing with another invisible enemy—loneliness? If he didn't have social support—not just from me but also from our families and friends—through his surgery and treatment, I'm honestly not sure whether he could've made it to the other side.

Yet I wondered how people who weren't lucky enough to be in healthy relationships could avoid the many risks of social isolation. Could single people, or unhappily coupled people, or really anyone who's experiencing loneliness, find a way to defend themselves?

Yes, but to fight loneliness one has to be willing to identify themself as lonely. If you constantly feel dissatisfied and unfulfilled by your relationship and your social life, if you feel like you're missing out on companionship, don't just dismiss it. That feeling is dangerous.

The next thing you have to do is *not* to trust your lonely mind. As Wilson reminds us, loneliness can play tricks to counteract our feelings of social isolation. Yet loneliness can also make us avoid or scrutinize the very social contact that we're starved for. So price that into your social expectations. And understand that when you're lonely, you might be less inclined to give someone the benefit of the doubt. You also might underestimate how much you'd benefit from a social connection.

A colleague at the University of Chicago, the psychologist Nicholas Epley, has found that people significantly underestimate how much meaning and joy they would get from striking up a conversation with a stranger, and so they don't. You won't be surprised to discover that lonely people in particular are likely to underestimate how much any given social interaction will mean to them.

Before I met John, I had spent all my adult life as a single person. Even though I sometimes felt stigmatized for choosing to be alone, and received much well-meaning but insistent advice about how to "meet people," I don't think I ever truly felt lonely. That's because from the earliest age I didn't expect to end up with anyone, so I felt fulfilled by my social reality.

Objectively speaking, I was alone, but subjectively speaking I didn't feel isolated. Remember that the brain doesn't care about labels. What it benefits from, what it *needs*, is a rich reciprocal connection with someone else, or something else. It doesn't matter how many connections you have or how society expects them to look; what matters, ultimately, for your physical as well as psychological well-being is the quality of those social connections.

How to Fight Loneliness

It's very hard to come up with a prescription to ward off loneliness, but during the forced social isolation that came with the coronavirus pandemic, I was asked for one by everyone from CNN's Dr. Sanjay Gupta to my eighty-year-old female neighbor to professional athletes to random strangers on hiking trails. My suggestions boiled down to five letters that spell "GRACE."

G.R.A.C.E. is an acronym for how an individual can take care of their social body—even during periods of isolation, when the Love Network is powered down and we are particularly susceptible to the dangers of loneliness. G.R.A.C.E. stands for Gratitude, Reciprocity, Altruism, Choice, and Enjoyment. Let's tackle each of these in their turn.

Gratitude. Lonely people usually don't feel particularly grateful for their lot in life. Yet try to force yourself, as you do at the Thanksgiving table, to come up with things for which you're thankful. It could be your family or your dog or your health or the weather outside or even yourself (for making it through every day). Every day, try writing down five things that you truly appreciate. Studies show that such simple exercises can significantly improve subjective well-being and reduce feelings of loneliness.

Reciprocity. The worst thing you can do to a lonely person is try to help *them*. If you know somebody who's lonely, ask them to help *you*. Being shown respect, being depended upon, being made to understand your own importance—all these things can give a lonely person a sense of worth and belonging that decreases feelings of isolation. The psychologist Barbara Fredrickson suggests cultivating "micro-moments" when you can connect in a small way to another person—whether it's a family member or the cashier at the grocery store. By making opportunities to share just a little bit of yourself, you'll receive in return a mood boost and a bit of stress relief that can build up over time.

Altruism. Volunteer—at the library, the running club, the Red Cross, you name it. Be part of something bigger than yourself. Helping others, sharing your knowledge, feeling a sense of mission—all this will give you a feeling of self-expansion that is similar to what people experience when they're in a loving relationship. But be careful not to spread yourself too thin or give of yourself in a sporadic or ad hoc way. Whatever altruistic cause you devote your time to, you have to make that time commitment a *regular* part of your life. When people do this, the results are impressive. For example, in a study of 5,882 adults over the age of fifty, the sociologist Dawn Carr and her colleagues showed that volunteering more than two hours every week reduced feelings of loneliness among widows to the same level as married women in the sample.

Choice. Understand that even though it might not feel this way, being lonely is a decision. The same situation can feel extremely friendly to one person and isolating or ostracizing to another. It all depends on your mindset. You can decide right now—yes, right now—if you want to feel lonely or happy. When we look at psychological interventions for lonely people, changing their attitudes and outlook has more effect on their loneliness ratings than increasing opportunities for social contact. Also, the way you frame your social life can affect the way you experience it. A 2020 study from Harvard University found that when people were made to sit alone in a waiting room for ten minutes, they usually felt bored or lonely—but significantly less so if they simply thought about "the benefits of solitude."

Enjoyment. This might sound like the most obvious advice imaginable, but try your hardest to have fun. Enjoy life. Science shows that enjoyment is a predictor of well-being and life satisfaction. Luckily, positive events tend to occur more often than negative ones. Yet not everyone makes a point of enjoying them, a process psychologists call *capitalization*. Sharing good news and good times with others helps increase positive emotions and reduce loneliness. And an interesting study by the social psychologist Shelly L. Gable of the University of California at Santa Barbara suggests that, in close relationships, couples who make the time to enjoy life together and share good news with each other are happier together.

10

The Test of Time

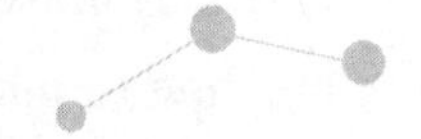

Light tomorrow with today.

—ELIZABETH BARRETT BROWNING

Against all odds, the Trilogy was working. The combination of top-notch medical treatment, John's fighting spirit, and the protective power of love helped him beat back the cancer. But the intensity of that fight nearly killed him.

The first blast of chemo and radiation lasted fourteen weeks: one week on, one week off. John would get hit with a wave of chemicals and photon beams, then we would spend a week home alone, waiting for the inevitable crash. After the first week at the hospital, we realized he had overdosed on chemotherapy. I remember calling the nurse constantly, asking if his extreme side effects—high fever, vomiting, an awful case of oral thrush—were normal. Soon he couldn't keep down any food. The doctors wanted to place a gastrostomy tube in his stomach so they could give him nutrients directly. He had been in such great shape before he got sick—his belly nothing but skin and tightly packed muscle—so after the doctor made the incision in his abdomen to insert the feeding tube, the pain was so intense that John woke up (despite anesthesia) and started screaming in the recovery room. He was terrified, delirious, thinking that such pain could only have been the result of a gunshot wound. He muttered something about "needing to protect Obama" (who was not only president at the time but also a living legend in Chicago).

When John came to, we laughed so hard about his Secret Service fantasy. And we needed that moment of catharsis because the news was going from bad to worse. The blood tests and scans showed all his markers heading in the wrong direction. The cancer was winning. We tried to stay resilient in the face of suffering, but after a while suffering gets old. We searched for hope in data. We couldn't help ourselves—we wanted to see the numbers, the deltas, the trend lines—even if they frightened us. We wanted to treat this like scientists, as if it were just another problem that could be solved through intellect and persistence.

While in the hospital John even tried to talk the doctors into tweaking their orders. He thought he might have a better idea of how to administer his anti-pain medication, given through a fentanyl patch, so he would not need higher and higher doses of the addictive opioid. He convinced the doctors to try it his way, but a day after they lowered his dose, he called them back, writhing in pain.

"So . . . my hypothesis was wrong."

Intellectually, I knew the chemo and radiation were taking their toll on his body, withering him from within, but when I looked at him, I didn't see a sick man, I saw my husband, the same eyes I fell in love with. And I felt in my bones that he would beat this. I genuinely thought he could do anything that he set his mind to. So part of me wasn't surprised when, after months of pain and uncertainty, things finally started breaking our way: the cancer markers improved, the scans got clearer and clearer. We celebrated each and every piece of good news with a "Happy Honeymoon" toast. By the fourteen-week mark, John's tumors had entirely vanished. He was in remission.

He returned to work, taking on a full course load the next semester, and restarting his research with renewed vigor. Yet we were not quite the same. The specter of death had taken residence in our apartment, taken a seat at the dinner table, like some uninvited guest. John knew his type of cancer had a high risk of coming back. He prided himself on facing up to the things other people ignored. Again, he said we should expect the best from life, but plan for the worst. He made it his business to "build memories," to create things for me to hold on to, just in case. We renewed our vows, with him wearing a tailored tux and me in a hand-embroidered white lace dress—the kind we hadn't had time to get before our first wedding in Paris. We drove from Chicago all the way down to his native West Texas, to see the small town where John grew up. That road trip had two purposes: He wanted me to see where he came from, to understand all the sides of him. And he wanted, literally and figuratively, to feel once again like he was in the driver's seat of his life.

After cancer, John became more selective about how he used his time. At work, he was focused on the *impact* of his research. He stopped advising graduate students and postdocs, focusing only on undergrads who had not yet chosen a path. "If you mentor grad students," he said, "you can change someone's career. If you mentor undergrads, you can change someone's life." He had taught yoga as a young man, and never had difficulty being "in the moment," yet I never saw him more focused, more present, than in those months after his treatment. In both his personal and professional life, his attention was on the *quality* of his relationships.

This probably came as no surprise to Laura Carstensen, the Stanford psychologist who had married us in Paris. She had spent years studying how people understood quality of life over their life span. Her theory of socio-emotional selectivity held that the way people understand their own life satisfaction tends to change as they age or face life-threatening illnesses.

Earlier in life, people perceive their future as "expansive and open-ended," and they are able to push thoughts of mortality to the margins. Thinking they have time to spare, people go into "collection mode," trying to accumulate things—money, status, knowledge—with an eye toward future use. Yet once people age or have a health scare, the internal calculus changes. Now they seek "emotional balance," they focus more on important and satisfying relationships and experiences, and become more interested in the present than the future, more interested in quality than quantity. The collector is replaced by the experiencer.

Once people make this transformation, they are usually better off. This clashes with the view that many young people have of old age, that the onset of illness and infirmity, the proximity of "the end," will bring about a sense of despair and depression. On the contrary. Not only are older people happier, their memory gets rewired to favor the storage of positive information. In one of Laura's studies, participants of various ages were shown a number of positive, negative, and neutral images. The young people were just as good at remembering images regardless of their emotional charge. But the older people remembered far more positive images than negative or neutral ones.

And we see this bias play out in relationships, too. As a couple age, they tend to focus on the positive, becoming more forgiving of the other person. It might sound paradoxical for a young couple, but perhaps one of the ways they could be happier together is to try to think more like older—or rather wiser—people, and challenge themselves to experience more and to take greater satisfaction in life. They might find that in doing so they focus on the positive side of life and let more negative stuff slide.

Beating the Clock

While some aspects of our biology help love survive the twists and turns of life, there are other hazards waiting for us, other ways that love will challenge us over the grand sweep of time.

The future, in general, is a scary place for people in relationships. This is because the vast majority of couples will sadly break up. We used to say half of marriages in the United States ended in divorce. But with fewer people getting married—and those who do, marrying later in life—the likelihood of divorce has fallen, to around 39 percent. That is still remarkably high. And the chance of an unwed couple breaking up is even higher.

The Stanford sociologist Michael Rosenfeld has been tracking breakup rates among gay and straight unmarried couples over time and found that, in young relationships, the likelihood of a breakup in the first year together is over 70 percent for both straight and gay couples. In the fifth year together, it falls to about 20 percent. From that point on, the risk of a breakup in any given year descends until you hit the two-decade mark, when it plateaus at around roughly 5 percent for gay couples and 10 percent for straight ones.

While the reasons why people break up are, of course, complicated, they usually boil down to two main problems: a lack of intimacy or connection with our partner (what we term *social reward*) or feeling rejected or unwanted by our partner (*social threat*). Of these two main breakup forces, psychologists believe that the lack of social reward is more decisive when it comes to a relationship's survival.

Whether it's a heartfelt declaration of love or a sweet smile, there are many gestures in a relationship that feel like rewards to the brain and body. And the love network thrives on reward; it needs dopamine like a plant needs water or like an electric car needs . . . electricity. Without it, your relationship can only go downhill. This is the reason why many breakups can be explained in terms of brain chemistry. As you'll remember from earlier chapters, when we fall in love, the reward system gets a big hit of dopamine, resulting in a blissed-out state of mind. Yet even in the most passionate couples, these early, intense feelings might change. Some couples progress into a longer-term, more emotionally grounded commitment, or they go their separate ways.

You probably have heard about the "seven-year itch," but a more scientifically grounded concept would be the two-year slump. For many people in relationships, the besotted brain—drunk on dopamine—falls down to earth at some point in those first two years.

Another dangerous period comes around the four-year mark. Helen Fisher has analyzed divorce rates across cultures and found that breakups tend to spike at this point in a relationship, which, she maintains, happens to coincide with the period of time when children need the most care to improve their chances of survival.

Whether it's the result of a deep-seated genetic tendency or a craving for dopamine, this natural "falling to earth" sensation after the first few years of a love relationship is exceedingly normal—and is not always a signal that your relationship is in trouble, just that it's changing or, if you prefer, evolving. Yet the absence of those positive, intense love feelings can make some people who have grown used to them feel like something is missing.

Justin Lavner, a psychologist at the University of Georgia, and his colleagues tracked 338 spouses through their first eighteen months of marriage and discovered that—regardless of the age of the couple or whether they lived together before getting hitched—most newlyweds experienced significant changes in their mood and personality over that short period of time. They found that couples became less agreeable, that husbands became less extroverted, and that wives became less open-minded. The reasons for this are, of course, complex, but a massive stumbling block for any couple has to do with not getting "stuck" in the relationship you had when you met and fell in love. We have to remember that just because there are tendencies for couples to pull apart from each other after the first few years, there is a difference between a tendency and destiny.

To Thine Own *Selves* Be True

Has anyone ever told you, "Just be yourself"? I can think of no piece of social advice that's more popular—or less instructive. When someone says "Just be yourself" to me, I usually reply, "OK. Which one?"

As far as the brain is concerned, the self is fluid. There is no one settled "self," and in fact separate regions give us sometimes conflicting information about who we are. Parts of the brain's PFC host and store self-conscious *labels*, our personality traits—whether we think of ourselves as generous or sophisticated or tall or handsome. The angular gyrus and other parts of the parietal lobe, meanwhile, foster a more symbolic, less superficial idea of the self. The PFC tells who you are *on paper*. The angular gyrus is more in tune with who you are deep down based on what you feel, do, and experience—what is sometimes called your "true self," which is often hard to put into words.

My point here is that, from the brain's perspective, the self is not a static thing. The self is dynamic, ever-evolving, and composed of many elements. And since relationships are composed of two selves, they're doubly so. We forget this too often. The trick to remaining satisfied and happily in love is to never forget that you and your beloved are always evolving, and so you always have to be engaged in the process of "getting to know" your partner, even if you have been together for a half century or more.

So don't worry about "being yourself"—as we've learned, that is a moving target—and instead be authentic to who you are at this moment. And then your authentic self will shine through—and your relationship will be stronger for it. Just look at John and me.

All through the cancer battle and the uneasy truce that followed, our bond never frayed. We got closer, in fact. What made our relationship adapt, what made it stay strong, is that we did not pretend everything was the same, we did not dwell on the way we were, we refused to live in the past. Instead, we were alive to all the ways that the present was challenging us, changing us, forcing us to evolve into a more united self. We talked about everything, we realigned, we found our footing once again, even though the ground beneath us was shifting. No matter what state John was in, I always treated him like my husband: I shared my joys, my fears, I asked him for his advice, I asked him for help.

From our research, both of us understood the importance of *reciprocity* and *mutual aid* in the evolution of social species. If I had treated John merely as a patient, pitying him, he would have lost the sense of his purpose within our relationship—an awful, disorienting experience that would make him more susceptible to feelings of loneliness. Instead, we kept our bond strong by practicing what psychologists call *self-disclosure,* by sharing the kinds of invisible truths that we all carry around in good times and bad—hopes, joys, memories, the sundry stuff out of which we build and rebuild ourselves, day after day.

Love Is a Choice

One of the best-known venues for discussions of contemporary romance in all its dizzying variety is the *New York Times'* "Modern Love" column, which has been published every week since 2004. Given the volume of submissions they receive (eight thousand each year), the column represents some of the most coveted real estate in the whole newspaper. And in 2017, for a special anniversary issue of "Modern Love," John and I were honored and humbled that the newspaper decided to profile us.

We related the story of how we met, how our recent brush with death had only brought us closer; I boiled down some insights from my research; a photographer came to our office and took pictures of us laughing in lab coats. When the paper came out, John and I bought out every newsstand in our neighborhood so I could send copies of the article to my family members in Europe. We received dozens of messages from our friends and colleagues. It felt like our love story was now official in some way, and we hoped that it would inspire other people. While the article made us very happy, its impact did not compare to the viral popularity of a "Modern Love" essay published two years earlier, titled "To Fall in Love with Anyone, Do This."

That essay was submitted by Mandy Len Catron, a writing instructor from Vancouver, who shared a story about replicating a famous social psychology experiment with one of her good friends. The experiment had been first carried out by Arthur and Elaine Aron and several of their colleagues in the 1990s. It tested whether you could, by hijacking certain features of human nature, make two random strangers fall in love. The researchers sent a heterosexual man and woman of similar age into their lab through separate doors. The participants sat down together, faced each other, and took turns answering a series of thirty-six increasingly personal questions.

The questions started out rather banal (*Who's your ideal dinner guest?*) but escalated quickly (*Do you have a secret hunch about how you will die?*). Soon the questions prompted the participants to engage with each other in different ways. *Make three true "we" statements each. For instance, "We are both in this room feeling . . ."* The questionnaire was specifically designed to elicit *self-disclosure*. Not only did the Arons and their colleagues find that their experiment quickly and reliably created the kind of intimacy essential to the formation of romantic love, they also had an impressive anecdotal piece of data: Six months after their experiment, two of the research subjects got married, and they invited the whole lab to the wedding.

Like any good scientist, Catron noted the limitations in her adapted version of this experiment: she was in a crowded bar (where presumably she was drinking) and not in a laboratory; her partner was a friend, not a stranger. Yet as they worked through the questionnaire, she was surprised to feel a deep well of affection opening up between the two of them that had never existed before. The experiment famously ends with four minutes of staring directly into your partner's eyes without speaking. Catron wrote that this experience was more thrilling, more frightening, than anything else she has ever done, including dangling by a rope from a rock face. In the first two minutes, she could hardly breathe. But then she began to relax, then it began to feel less awkward, then it began to feel *right*.

The experiment hadn't put her under some spell. But rather, she said, it showed her that it was possible "to generate trust and intimacy"—the very feelings that underlie love—through action. And, in the weeks and months that passed, she and her friend actually did fall in love. She concluded her beautiful piece this way: "Love didn't happen to us. We're in love because we each made the choice to be."

For those searching for love, for those who are not currently satisfied with their relationship, this idea of choice is empowering: It means that if you're not satisfied, you can do something about it. And once you and your partner begin to self-disclose, one of two things will happen: (1) your relationship will get stronger, or (2) you will realize that you shouldn't be in your relationship.

Studies have found that, as self-disclosure increases in relationships, so does relationship satisfaction. But it's a two-way street. Couples who complain about low intimacy also tend to self-disclose less. And a lack of self-disclosure makes you and your partner vulnerable to all the perils of loneliness that we discussed earlier, such as a greater susceptibility to disease and early death.

When the German psychologist Marcus Mund and his colleagues examined feelings of loneliness in nearly five hundred long-term couples, they found no association between loneliness and the quality of a physical connection in a relationship. But they found a strong association between loneliness and a lack of self-disclosure. This means that couples who don't reveal their "true selves" may be setting themselves up for long-term suffering.

The Code to Heartbreak

Sometimes, as much as we might try to hold on to love, we lose it. There can be a breakup because of some irreconcilable difference, a family conflict, relocation, differing priorities, a lack of physical intimacy, you name it. When we are separated from a person we once loved, what happens inside the brain?

Nothing pretty. After an unwanted breakup, the parts of our brain responsible for craving rewarding sensations become overactive. You love even *harder* than when you had your partner, searching for the person who's no longer there, for the positive feelings you once associated with your beloved. This is what heartbreak or unrequited love looks like.

In addition to the reward system, the part of the PFC that is involved in rumination is also activated. This is the part of the brain that controls our tendency to think over and over—and over again—about our partner and our relationship, what went right, what went wrong. Finally, we see activity in brain areas that respond to pain, like the anterior cingulate cortex. In brain imaging studies of women who had recently broken up with a loved one, when they thought about their partner, they activated the same regions that are triggered when we experience the grief from the death of someone close to us. That is, from the brain's perspective, the difference between a bad breakup and a death is negligible.

As neuroscientists get better at pinpointing the way love and heartbreak work in the brain, we might not be far away from a future in which "getting over" a bad breakup involves not only a box of tissues but also an EEG machine.

Consider the case of the multitalented musician and writer known as Dessa. As a widely published author and a member of the popular hip-hop collective Doomtree, she had achieved an enviable level of artistic success. Yet she had not been as lucky in her love life. As hard as she tried, she could not get over an ex—a guy who had been in and out of her life for over a decade, a person she couldn't live with or without, who triggered in her a cascade of positive, negative, and confusing feelings—frustration, regret, jealousy, desire. She did the math; she figured out that, at the end of the day, he was bad for her. Yet she could not shake free of him. "I was not only heartbroken, but also embarrassed that I couldn't rebound . . . I couldn't figure out how to put the love *down*."

She decided to do something about it. She read the latest scientific research on love. She found out that neuroscientists were able to use brain imaging techniques to measure and localize feelings of love in the brain. She wondered whether this science could also be used to zero in on those feelings and retrain her brain to "fall out of love" with her ex.

On Twitter, she found a professor of neuroscience at the University of Minnesota, Cheryl Olman, who agreed to scan Dessa's brain in the fMRI machine while showing her images of her ex-boyfriend and a stranger who shared some of his features (the control). When they compared the fMRI signatures, several of the key love "reward" areas lit up for the ex (including the caudate nucleus and the ventral tegmental area), as well as the anterior cingulate, which registers feelings of pain. Dr. Olman sent Dessa an image of her brain in cross section, showing exactly where her feelings for her ex resided. Looking at the scans, she was determined to find a new way of thinking about this old love.

For that task, she turned to a method called neurofeedback, also known as EEG (electroencephalogram) biofeedback. In brief, neurofeedback is a tool that measures brain waves and displays the results live to the person wearing the headband, so they can become aware of what their brain is doing at certain moments. The point is to learn gradually how to train your attention and regulate your emotions in specific situations. The therapy uses various output signals—like sounds or visual markers—to help you reorganize or retrain your brain.

After subjecting herself to nine neurofeedback sessions, Dessa felt herself fixating less on her ex, as though he loomed smaller in her emotional world. And a subsequent fMRI scan in Dr. Olman's lab showed that, when she was shown a photo of her ex, the brain regions that had previously been hyperactivated were now quiet. Whether it was because the neurotherapy helped, or because she had begun through this exercise to think and talk about her heartbreak in an entirely new way, Dessa had finally managed, after years, to find a path beyond the heartbreak. And that path went straight through the brain.

11
Shipwrecked

How was I able to live alone before, my little everything?
Without you I lack self-confidence, passion for work, and enjoyment of life
—in short, without you, my life is no life.

—ALBERT EINSTEIN

John and I were as connected as a couple could be. We had fallen in love despite an ocean between us; we had dissolved our differences in culture, language, and age; we had faced stage IV cancer, and beat it, together. We not only *felt* like our love made us stronger and smarter, we had generated reams of data to prove it. Yet at some point I must have drunk too much of my own Kool-Aid. I was walking around thinking of love like a superpower, thinking that if we just loved hard enough we could survive anything. As a scientist, I should've known better.

In the two years since John was diagnosed, he nearly died so many times that I lost count. Getting that close to the end made us fearless, in a way. It clarified something, it taught us how to live. The things that were really important before John got sick—work, exercise, taking care of family and friends—became absolutely vital to us; the things that were less important—the gray in our hair, the cold weather, the endless traffic jams, the number of likes on social media—became almost irrelevant. The color of life changed, intensified; leaves never looked greener; our time together seemed sweeter because we knew it wasn't guaranteed to last. As people who love learning, who have the freakish ability to keep our eyes wide open no matter what is happening, good or bad, we couldn't help but find this new keener life interesting, at times even sublime. In many respects, as we got closer to the edge, we enjoyed the view. Maybe the simple fact that I wasn't living in fear of what *might* happen explains why I was so unprepared for it when it finally did.

It is hard for me to write this chapter. It is hard because to think about all this is to relive it in a way. As I write these words, I'm not sure I want to keep them on the page. I'm not sure I want to go back to this time. Yet there is no way I can finish my story, no way I can really find out the true depth of love, without telling you what it was like to lose it.

I don't remember much about that night—shards of images that I've pieced together. I will tell you that what happened came as a great shock to me because John had recently been getting better. Though his cancer returned and had spread to his lungs, he had fought back. After his first experience with the Trilogy, he had been so weak he couldn't lift a hanger in our closet, but he forced himself to go back to the gym every day, and built back his body piece by piece, starting with five-pound weights, then ten, then forty. By the fall of 2017, he was ready for a victory lap. He had regained his appetite, regained his hunger for research, he had never looked better. Many people who met him during this time had no idea he was even sick. He received one of the highest honors from the University of Chicago, the Phoenix Prize, as well as a medal from the Centers for Disease Control in Washington, DC, for his work alerting the world to the danger of loneliness, which—thanks to him—was now being treated like a full-blown health epidemic.

We toasted the new year feeling hopeful and content. But in the first few weeks of 2018, John's medical news went from good to bad to worse. A series of medical complications related to the cancer treatment required us to stay in the hospital for several weeks. One night his numbers went so low that the nurses and doctors thought his day had come. They asked me to say goodbye. But, miraculously, his vitals normalized by the morning. He was discharged in February. We went home, and his condition improved, except for a nagging cough. Friends and neighbors took turns bringing us food and walking our dog—so we could spend time together focusing on John's recovery and adjusting to a new normal.

On March 5, John had a follow-up outpatient treatment at the hospital. The doctors said that he had "turned another corner." We relished that moment—I remember John on the phone with our friends, the huge smile on his face as he talked about the good news. We went to bed that night with a sense of relief.

But two hours later John started coughing harder than ever before. He seemed unable to catch his breath. And then, in one awful instant, he felt something inside him give way. His mouth filled with blood and he instantly knew that this was the end. He had just enough time before losing consciousness to look at me and say, "I love you."

I called 911 and administered CPR. When the paramedics came, they tried to resuscitate him for several minutes before stopping.

"Please," I told the female paramedic. "Please try again."

There was no medical possibility that he could be brought back, but she tried once more, for me. When she told me he was gone, I refused to believe it. I kneeled and begged her in tears to let me try one more time. All the paramedics looked at one another—and in silence one of them gave me the "yes" by nodding her head up and down. I tried CPR again, until at a certain point I realized what was happening, what *had* happened, and started screaming.

Not Letting Go

I was in shock. I couldn't process what the paramedics meant when they said they were going to take John away. But once that idea sank in, I told them that I had to go with him. I couldn't be separated from him. We took the elevator. It was the same thirty-flight trip down we had taken thousands of times before. Now it would be the last time.

As the doors opened on the lobby, I was following John, who was lying on the stretcher. I could feel the eyes of the security guards and doormen on us. The paramedics wheeled John toward the waiting ambulance. It was in this moment that I experienced the first flash of understanding that our life, my life, had suddenly changed. The feeling must have sparked something in my angular gyrus, because I experienced a temporary out-of-body sensation, where I saw this moment from above—it felt like a protective measure, a way of the mind dissociating slightly from the body to get some safe distance from the painful reality that I was living through.

And that reality was refracted through the tearful eyes of our neighbors, our friends, and the people in the building who had taken care of us. I could feel through my mirror neurons their suffering, the collective sigh of anguish, which made me want to crumple.

I was excruciatingly sad not only for John but also for them.

We were "the couple," always together, always smiling, and now *we* were no more. These people were the only family I had in Chicago. A few of my female neighbors came with me to the hospital morgue. As we spoke about the arrangements, I was there but I was not there. At a certain point, I stood up and told the funeral home director that I needed to see John again.

"We don't recommend that," she told me, explaining how the body begins to change, how John wouldn't look like himself. I couldn't care less. I *needed* to be next to him. I *needed* to be next to my husband. They brought me to the room where he lay, and I cried and talked to him. When the director came in and kindly suggested that I finally leave, I leaned over John, and kissed him, and once again told him that I loved him.

I felt like this was all wrong, like we shouldn't be there, like we needed to go *home*.

John and I had talked about everything, so of course we had talked about what I should do, and shouldn't do, if he died: what to tell our friends, our colleagues, the press; what I should do with my job and our home. Yet we had never talked about funeral arrangements. I became fixated on this idea of "going home." We would do the funeral in our house, among our things.

In a French-Italian family like the kind I grew up in, the idea of having a funeral at home is not as strange as it might seem in America. In my grandmother's native village in Italy, for instance, when someone died they were kept in an open casket at home and friends and neighbors came over to visit, pay their respects. The family often decorated the door and publicized the death in the village, so people could share in the mourning process and offer support at a time of need. In my family, widows dressed in black for a year; we had a protocol to follow, a way of behaving, a channel for our grief.

I felt close to this tradition and I think John would've wanted it, too. But a big part of why I wanted to bring him home was that on some level I was not willing to believe what had happened. Part of me was in denial. Part of me thought somehow this was all temporary. I felt like I was fumbling for the switch that would bring us back to the world we knew before. Maybe he was just away, in the hospital for a few nights, for some new procedure, and if I were resilient and focused and stubborn, I could find my way back to him.

The Walking Dead

I felt fortunate that my best friend and next-door neighbor, Fernanda, who happened to be a clinical psychologist, could visit with me every day. She knew that in this moment of intense crisis all she could do for me was be there and sit in silence as the tsunami of feelings crashed over me. And every time I'd thank her for this extraordinary favor, she'd simply reply: "It is not a favor—it is love."

I was also fortunate to have a friendly relationship with a number of Orthodox Jewish women who lived in our building. Despite caring for big, growing families, these women rarely seemed hurried. They were always so calm and welcoming, always ready for a little chitchat, to make jokes about our dog or their kids or their grandkids or the weather. They heard my screams the night John died. They immediately rushed to my aid and knew what to do. They all held me while I was holding John's hand, unable to let him go.

In the following days, the women in this community adopted me, even though I didn't share their faith, even though we were only neighbors. I accepted their comfort, their kindness, their matzo ball soup. And they taught me the Jewish tradition of shiva, which gave a kind of structure and protocol to my grief. I grieved in a holistic way—applying the traditions from both my Catholic ancestors and my Jewish neighbors. I dressed in black. I covered the mirrors in my home. I wore a torn black ribbon that symbolized my loss. I listened to a diversity of soothing prayers from many different religions. I felt I had been given the space in which to share my sorrow.

Although people sometimes use the words "grief" and "mourning" interchangeably, scientists and mental health professionals think of these things as separate, if related, concepts. Grief encompasses the thoughts and feelings you experience after a loss; mourning is how those internal states are expressed externally. In some traditional cultures, the mourning process can be ritualized and even rule-bound. In China, the color red symbolizes happiness, joy, and luck, and is one of the traditional colors worn by Chinese grooms and brides, so people don't wear that color after someone dies. In the Philippines, a vigil with an open casket is often held for up to a week, during which one is prohibited from sweeping the floor.

Other than a few stable traditions (like the funeral ceremony), modern-day mourning in the West looks very different from person to person. There is no set rule book for one to follow. The benefit of this is that people can mourn a loss however they see fit without the pressure of observing the "proper" way to mourn. The downside is that if they don't know how to mourn or honor their loved ones, they might feel—in addition to the intense pain—a sense of social confusion, a helplessness, a lack of direction.

I needed a protocol. Shiva was the closest thing I could find to my grandmother's ritualized style of mourning. I needed a guardrail in my mind to contain and control the excruciating pain and the chaos. My neighbors made sure I wasn't alone. They kept me alive, but I was only barely alive. I couldn't even leave the building to walk my dog—neighbors had to step in to help. The first time I was able to make it down to the lobby, all the staff in our building saw me and immediately came over. They gave me a group hug, like a basketball team coming together after a game, only we were all crying, united in grief.

For weeks afterward, I wore John's baggy hoodie over my small shoulders and covered my head in a baseball hat that I had given him. The hat bore the letters *RF*—for Roger Federer. We were both big tennis fans, but over the years we had decided to change what those two letters stood for to "Romantic Forever." That cap, his hoodie—they became my uniform, my second skin.

The weeks turned into months. The outpouring of warmth and kindness that I had received in the aftermath of John's death slowed to a trickle. People in the building that I was less close to started avoiding me. They had already expressed their condolences, and now they didn't know quite what to say. I found myself turning inward. I didn't want to be seen. I was getting a bit tired of the pitying glances. I hid myself under my hat and my sunglasses and my big sweater. Soon people stopped recognizing me, or pretended to. They were treating me like the ghost I had become.

All my knowledge of the emotional brain, all my knowledge of human psychology, seemed inaccessible, almost meaningless, in the immediate aftermath. I couldn't manage doing anything for myself. I couldn't even find the motivation to make a cup of coffee. I felt helpless. And yet, a few weeks after John's death, I had to organize a memorial service for him. I never could have put such an emotional event together without the kindness, support, and guidance of our families, friends, neighbors, and colleagues, including the then president of the University of Chicago, Bob Zimmer, and his wife, Professor Shadi Bartsch-Zimmer. The memorial took place at the historic Rockefeller Memorial Chapel at the University of Chicago, where John had given a convocation speech.

The chapel was now decorated with white flowers generously

sent by the crown princess of Denmark, who knew John well and was inspired by his work on loneliness, which she made a priority of her pioneering nonprofit foundation. The university flew the American flag at half-staff, something I had never seen for other professors who had passed away. The university's bagpipes master played "Amazing Grace." I came veiled in black. I remember barely being able to hold a conversation with anyone. Dr. Jack Rowe was there, the Columbia professor who had given me away seven years before at our wedding in Paris. The only word he knew to describe the look on my face came from his reading in German philosophy, *Scheitern*. Loosely translated, it means "shipwrecked." That is exactly how I felt, like a once-seaworthy vessel now sinking fast.

I held it together during my eulogy. I knew I wouldn't have been able to talk about what John meant to me without breaking down, so I focused on everyone else who was there and who wrote to us from around the world. I knew I could speak for him in expressing gratitude, in reaching out to our family, friends, colleagues, and students and thanking them for all their support and concern. I ended my short speech by thanking John himself. Talking to him at this very moment gave me a (false but soothing) sense that he was still present with us—a feeling I needed in order to have the strength to stand there. I said I was grateful to him for falling in love—with science. I talked about his research and his brilliance, about how he paved the way for a new understanding of social connections, giving us empirical evidence that a more meaningful life was a life connected to others. And yet as I looked out on that sea of weeping faces, I knew that the connection that had mattered the most to me had been ruptured, and I honestly wasn't sure that, for me, a meaningful life was still possible.

12

How to Love a Ghost

The worst thing you can tell a person who's grieving is time will heal.

—JOHN T. CACIOPPO

I was grieving, I was lonely. Fortunately, I had been married to a scholar of loneliness, who knew how to overcome grief, who left reminders of what to do, how to act, everywhere I turned. One of the things I came back to over and over again was a talk that John once gave to seniors at an AARP convention. The talk was all about how to care for someone who had lost a loved one. I found it on the day of John's funeral, as I had been looking for a quote to share with our family and friends, a kernel of wisdom that could console them and guide them through their pain. I typed his name into YouTube, found a video I had never seen before, and then pressed "play." All of a sudden, he was speaking directly to me, his big puppy-dog eyes looked watery, even kinder and more empathic than usual, as if he could feel the pain of the people he was talking about at that moment.

John was describing a longitudinal study of loneliness that he had started and we were still conducting with elderly people in Chicago. It was then in its eleventh year, and many of the people in the study had recently endured major losses, such as the death of a best friend or a spouse of fifty years. John understood that for these people, such a loss felt "like the end of the world." But he saw, time and time again, how study participants had "risen from really crushing episodes of social isolation." He urged patience. "Sometimes when the world looks darkest, we need to turn that adversity to an advantage, we need to figure out what are the opportunities that are now available, and not give up." Yet, as always, he had a critique of the conventional wisdom. "The worst thing you can tell a person who's grieving is time will heal. It's not time—it's the actions, cognitions, how you approach other people."

In the following months, I would think a lot about this idea, that it's not time that heals grief but *other people*. I meditated on it like a koan, and sometimes I argued with it. Because the truth is, as much as I remained a believer in the wisdom of our research, I wasn't remotely interested in other people after losing John. I wanted only *him*.

How, John, how could I let new people into my life now? *How can I reach out to someone else, when I'm faced with so much grief, when my brain is sending distress signals, when the love network is powered down, when my angular gyrus—the part of me that grew and expanded to include you—has gone dark, making me feel not only like I've lost my husband but also like I've lost myself?*

But there was no winning an argument with John, even beyond the grave. I could hear his voice, now playing in my head, that mix of warmth and cool rationality, telling me that even though he was gone our love remained biologically encrypted in my mind. What he didn't tell me, or what I was unwilling to hear, was that to reactivate the love network required the strength to face the sadness, the pain, that comes with the loss of a partner.

Good Grief

I was shocked by how much John's death *hurt*, not just psychologically but physically. My heart literally burned for weeks, I could barely eat, I lost twenty pounds in a month. Living through the death of someone you love is one of the greatest stresses human beings can experience, and that stress is felt keenly in the body. This explains why surviving spouses suffer so many serious medical events during the bereavement period. Your resting heart rate increases, as does your blood pressure. Your body is flooded with the stress hormone cortisol, and your immune system is depressed. In rare cases, the shocking news that your loved one has died can itself be lethal.

In the twenty-four-hour period after a loss, a person has between a twenty-one to twenty-eight times higher risk than normal of suffering a heart attack, depending on how close you are to your loved one. And even if they don't have a heart attack, some people think they are experiencing cardiac arrest when they're actually suffering from "broken heart syndrome," a rare condition in which acute stress causes extremely painful changes in the shape of the heart's main pumping chamber. So yes, in rare cases, you can literally die from a broken heart.

Yet even if you survive the initial shock of the death of a loved one, you remain at risk for months after. One groundbreaking study of grief from the 1960s looked at 4,486 widowers in Britain. In the six-month period after they lost their spouses, they had a 40 percent higher risk of dying than a married person their own age. After this period, their mortality rates began to track with others in their age group. Yet more recent studies have shown that people who suffer the death of a significant other, especially if they remain tortured by their grief, have an elevated risk of developing cardiovascular disease, diabetes, and cancer long after the period of acute grieving is over.

While grief hurts our bodies, it torments our brains. When you are grieving, you can't think well. The brain's alarm center, the amygdala, is overactive, while the "regulating and planning" center of the brain, the PFC, is underactive. That's why people may have trouble doing simple tasks—they are lost in a fog of grief. They might forget to exercise, to eat, to put the coffee in the machine. They drive straight past their exit on the highway.

Part of the reason we're so distracted when we grieve is that we're thinking about the loss not only from our own perspective but also from the point of view of our lost loved one. Remember all that we learned about the mirror neuron system. That empathetic response that we felt toward our partner when she or he was living remains intact even after our partner's death. When we see their photograph or visualize them in our mind's eye, we can't help but imagine what they would think and feel about their own demise. I know I did this for John. I knew I was the only person in our relationship who was still suffering, yet I focused intently on *his* suffering, as if he were still feeling it. I thought: *It's not fair*. And: *He's too young*. I wished over and over again that I could take his place.

This is part of what psychologists call "grief rumination." Whether you're torturing yourself with counterfactuals (What could you have done differently?) or dwelling on the injustice of it all (Why him? Why us?), you are in a way reliving the death over and over again—

visually and viscerally. And just as with an unwanted breakup, the brain areas involved in flashbacks and autobiographical memories turn on. A movie trailer version of your life together flashes before your eyes, and it always has a sad ending. The brain areas involved in somatic or bodily sensations also activate, which makes you *feel* your emotional pain in physical ways: tightness in your chest or limbs, shortness of breath, headaches, strange numb sensations.

All this psychosomatic danger pushes the brain's main threat detector into red alert. Even though you may feel like your worst fear has already been realized, the amygdala, the seat of our survival instincts, is firing on all cylinders, sending signals to the hypothalamus to release chemicals, putting our body in a constant fight-or-flight state. You can sustain this state for days, even weeks, but if it persists you are in trouble. As we know very well by now, we are not built to feel this way day in and day out. When we do, when a loss triggers a stress response that won't go away, it has the potential to rewire the brain's circuitry and fry the mind.

It's Complicated

Acute grief comes in all flavors: some feel anger, depression, hopelessness. Some people dissociate, act impulsively, repress. It is popular to speak about grief in "stages," which sometimes makes the process sound suspiciously like a recipe. *Five stages of grief—and it's done!* The truth is: For most people grief is a cyclone, a whirl in which you could feel many things at once or the same thing—over and over again. People only like to think about it in steps, hoping these steps will lead us to someplace . . . better.

For most people, this is true. Six to twelve months after the loss of a loved one, they emerge out of the fog of grief. They'll never be the same, exactly, but they begin to move on, to explore new options, to rise, as John put it, from those crushing periods of felt isolation. Yet about 10 percent of those who have lost a loved one still can't get over it after the first year. They are mired in what psychologists call "complicated grief." They have become lovelorn zombies, yearning for their beloved with an urgent sense of longing, even as they know intellectually that reuniting is impossible. Everywhere they look, they are reminded of the thing they cannot have. Such a state drains all the joy out of life.

The relationship between ordinary grief and complicated grief is similar to the relationship between ordinary loneliness and chronic loneliness. Both loneliness and grief are protective, evolutionarily adaptive biological signals. The first tells us we need to reconnect with others to survive; the second helps us deal with the trauma of loss. One has to learn to trust the process, to accept the brain changes that are occurring during grief, to pay attention to them, to use the urgency and strangeness of this period as an invitation to heal by embracing all the emotions that knock on your door. But some of us can't. And we slip into complicated grief, which just like chronic loneliness may be dangerous to our minds, our hearts, and our bodies.

Mary-Frances O'Connor, a psychiatrist at UCLA, and her colleagues showed people who were suffering from either complicated or uncomplicated grief photographs of their lost loved ones while scanning their brains. She found that a certain part of the brain's dopamine-driven reward system—the *nucleus accumbens*—was activated for people with complicated grief but *not* for people with ordinary grief. A neighbor of the amygdala in the most ancient part of the brain, the limbic system, the nucleus accumbens usually lights up when we long for something, when we search it out with an expectation that we will eventually get it. And neuroscientists have discovered that it is more sensitive to the *anticipation* of a reward than its *acquisition*.

A healthy kind of grief means that when you see a photograph of your lost love, you understand that it does not represent a "living reward" but rather the memory of someone who is gone. For whatever reason, complicated grievers can't wrap their brain around this fact. They are not coming to terms with the death of their beloved. On a deep level, their brain is still *expecting* to see them, to feel them again. Though the nucleus accumbens is located in the brain's so-called reward circuit, its hyperactivity is not a sign that good things are happening—quite the contrary. Complicated grief, if not treated, can be so harmful that some researchers have likened it to a traumatic brain injury and have found evidence that it may accelerate the onset of dementia and other forms of cognitive decline.

Breakdown

One of the ways people with complicated grief try to deal with their pain is by avoiding thinking about the person they lost. This makes sense—avoidance is a natural and even adaptive way to manage suffering. But here is yet another example of the way an adaptive mechanism can turn against us when we take it to extremes. Psychologists know that if people continue to avoid the emotions stirred up by grief, they will never overcome it. Eye-tracking studies have shown that people who ruminate more over the loss of their loved ones are also more inclined to *avoid* reminders of them. In the aggregate, such avoidance may take more mental energy than we would expend by facing and processing the emotions stirred up by grief, making us more anxious and less focused in other aspects of our life.

In my case, avoidance wasn't really an option. I was too focused on my grief to sweep it under the rug. John's absence was all around me. While I couldn't help but face it, that didn't make the pain hurt any less. In fact, dealing with that pain was the hardest thing I've ever done.

And I almost didn't make it.

Several weeks after I lost John, I was still crying myself to sleep, unable to take pleasure in any aspect of life. The memorial service didn't bring me closure—it only seemed to make his absence feel more real. I kept having flashbacks, seeing myself in a black veil leading the funeral cortege at the university chapel, breathing through the pain, walking toward . . . *what*? What now? I tried to go through the motions of social life: I made small talk with neighbors, I had coffee with friends. But I knew I was depressed—and I had never been depressed before. I was looking down all the time. I had no energy, no appetite. I walked by flowers that I could not smell. I saw the birds but was deaf to their song. I ate food that I could not taste.

I tried to break out of my shell of loneliness. One night, a month after losing John, a bunch of my neighbors gathered in our building to watch a basketball game. It was the Sacramento Kings against one of John's favorite teams, the Golden State Warriors. The room was packed, and everyone quieted as I walked in—they were surprised to see me, since I had basically spent the past four weeks in seclusion. But everyone smiled, happy to see me take this step toward "normal life."

In a way, I had come to love these people—the doormen who had helped us on the way to chemo, the women who sat shiva with me, the dog-walking friend who ran into me and Bacio the day after John died and knew, in the instant she saw my face, exactly what had happened. She had taken me in her arms and we cried together. All these people were there, having a good time, waving me over to join in their conversations. And do you know how I felt, surrounded by their caring faces?

Alone. Entirely alone. This, to me, was proof that life without John was no longer worth living. And it seemed to totally contradict his simple words of wisdom, about how "it is not time" but "others" that help us overcome grief. I went upstairs, opened the door to our apartment, collapsed on the floor. I was *done*, determined to end the suffering. I felt like my life had lost its meaning. My brain's alarm system, controlled by the amygdala, seemed to be on a self-destruct course, the blaring negative siren interfering with and drowning out the rational PFC, which typically would send inhibitory signals to calm down the amygdala.

I had hit rock bottom. Yet somehow, miraculously, at that moment, the scientist in me sprang into action. A skeptical hypothesis entered the maze of despair that had become my mind: If ending everything were a good idea today, it would still be a good idea tomorrow.

(For the record, it's never a good idea.)

So I decided to sleep on it. But before going to bed, I sent an SOS to an old friend who lived on the other side of the country. What I needed then was a kind of help that went beyond caring hugs and concerned looks and bowls of matzo ball soup from my kind neighbors, those things that were so essential in getting me through the first few weeks without John. What I needed now was for somebody to show me how I could help myself.

When I woke up in the morning, I could still see the abyss, but I felt in some respect that I was no longer leaning over the edge. The storm had passed. And there was a beam of light, an email from my friend, sitting in my inbox, an email that, in a way, would save my life.

My friend was a retired pro tennis player, someone I had met by chance years earlier. I thought of him as a little bit of a "mastermind" who knew how to keep calm in the most stressful situations. He was familiar with the first part of my story, how I had spent many years alone, how I thought I would never find love until I met John. Yet it had been a few years since we last had exchanged messages. He didn't know that John had gotten sick, much less that he had passed away.

After a few supportive email exchanges, we set up a time to talk on the phone. I'm not exactly sure *what* I expected him to say. I had been in touch with other friends and family, people who poured their hearts out to me, who tried desperately to help, to find the right advice that would shake me out of my grief. But nothing they did seemed to work.

My friend was able to tell after just a few minutes of talking to me that I was hanging on by a thread. Our phone call was brief, unsentimental. But he spoke the truth. He told me that if I could not trust my mind—the part of me that felt like I no longer even knew who I was—then I had to trust my body instead. He asked me if there was a park or someplace near my home where I could run. I told him about a two-mile track not far away.

"Good," he said. "Put your running shoes on and do three loops. Then call me at the same time tomorrow."

He didn't know how out of shape I was, how frail. I hadn't run six miles in ages. But like the good student I always was, I completed the assignment. Half a loop in, I was panting, sweating, limping—but I decided to keep going. I walked the remaining five miles. The next day I was so sore I felt like staying in bed all day, but my friend—whom I had taken to calling Coach—told me to run another six miles. I did that. Then another. Then another.

Every day, for one year, I ran six miles. Coach also told me what to eat, how to balance my fluids versus solids, even what I should read every evening before going to bed. He also sent me a list of inspirational videos and documentaries—stories of athletes who went through hell, who lost their families, lost their limbs, overcame dire poverty and abuse to become champions. Those stories fueled me in some way.

I checked in with Coach regularly. He didn't reply to every message. But every week I would hear from him. This intuitive technique to answer messages in a seemingly random fashion is what psychologists call—in their own nerdy way—"operant conditioning with variable-ratio schedule." The idea was to keep me guessing, to create unpredictable rewards that have the effect of turbocharging behavioral change while maintaining self-resilience in the subject.

It may seem counterintuitive, but sometimes what people need most in these moments is not a shoulder to cry on but a hand to hold or, in some cases, a swift kick in the rear end. Coach's approach was *tough* love. If I tried to impress him, he rebuffed me. When I told him I ran an eight-minute mile, he replied: "My grandmother runs faster than you." I remember running by the lake one day in winter as an icy rain mixed with hail pelted my face. It was almost unbearably cold, yet none of the physical pain I experienced in that subzero Chicago weather came close to the emotional pain that was waiting for me when I came home. In that moment, I wished I could have stayed outside running in the cold for days.

"Better to run toward," Coach told me, "than to run away." At first, I'll admit, I was just running away. But as I put more miles under my soles, I could feel the positive effects of the well-known runner hormones rushing through my body and mind—the endorphins, the dopamine, the serotonin. In the end, running would help me to simplify my complicated grief, to climb out of the darkness. I survived by capitalizing on both the natural function of the body and the natural resiliency and social nature of the brain. Just as John suggested, I found strength in other people (Coach, the stories of the athletes, the sense of inner fortitude they all embodied), but I also found it in myself. I was running toward my true self. After a few months, Coach told me it was time to return to a childhood passion, an old friend: tennis. Though I always preferred singles matches, I signed up for a women's doubles league. I was now ready for a partner, at least on the tennis court.

To Love You, After

There is one more love story I want to share with you: that of Richard Feynman and his first wife, Arline Greenbaum. Feynman is the third theoretical physicist I've mentioned in this book.

What is it about physics, I wonder, that produces so many inspiring romantics? In addition to mapping the trajectory of subatomic particles and winning the Nobel Prize in 1965, Feynman was a beautiful writer and science popularizer whose many books explained physics to the masses. Yet one of the greatest things he ever wrote was never published in his lifetime. It was a letter addressed to his dead wife, Arline Greenbaum.

Greenbaum had been Feynman's high school sweetheart. Though he went off to college and graduate school to pursue his illustrious career in physics, Feynman was always determined to marry his first love, and he became even more determined after she was diagnosed with a terminal case of tuberculosis. In 1941, they took a ferry from Manhattan to Staten Island and got married in secret, at City Hall. Two strangers served as witnesses. For fear of contracting tuberculosis, Feynman could only kiss his bride on the cheek. He wrote his letter to her almost two years after her death, and in it you can read the rational scientist pouring his heart out, trying, flailing, reaching to grasp the mystery, the *point*, of love after death. He opens up about how much she meant to him. He describes her as the "idea-woman," the "general instigator of all [their] wild adventures," and confesses that without her, he feels alone. He shares his inner fears, his hopes, and how much he'd love to continue to take care of her, to comfort her, to do "little projects" together—"making clothes" or "learning Chinese."

Perhaps even more than any of the scientific explanations in this book, Feynman comes closest in this letter to showing us what true eternal love means. He closes with a haunting, beautiful, startling couplet: "I love my wife. My wife is dead." Then he signs his name and adds a postscript: "Please excuse my not mailing this—but I don't know your new address."

My Last Lesson

If life is a roller coaster, then the people who will suffer the most are those who can't accept that they are strapped into a ride, those who can't come to terms with the fact that the ups and downs are beyond their control. I have found that, in the face of unrelenting fear, it is far better to open your eyes and scream, far better to cling to your friend's arms, or even to ask the stranger sitting beside you to hold your hand, than to try to control the uncontrollable.

I did not learn this in the laboratory or on the running track. I learned it by skydiving. This was the summer after John died. I was visiting Switzerland to spend time with family. For my birthday, some old friends decided to surprise me. They told me they would pick me up one morning, and asked only that I wear comfortable clothes and running shoes. I was excited, thinking we were going to hike to some scenic spot in the Alps. But when we arrived at the undisclosed location, I saw a number of tiny airplanes idling in a meadow and people wearing funny backpacks.

Then I realized those backpacks contained parachutes.

"Surprise!!!!" my friends screamed with big smiles.

The plan was for me to skydive alone with the instructor. My friends would wait for me on the ground, so they could take pictures. I was confused. I thought friendship was about sharing good times together, not watching your friend be terrified. *Don't they know I'm deathly afraid of flying?* Even though I have tried to heed the wisdom of the research to which I've devoted my life, to open myself up to the unexpected, to "let go," few things on earth scared me more than skydiving.

When the door of the plane opened, panic gripped me. The tandem instructor tried to explain to me that screaming when the doors of the plane open, and you feel that first burst of air in your face, helps people avoid feeling hypoxic and deal with the very natural and rational panic and fear that come from jumping out of a plane. Screaming helps your brain *accept* the pain, the discomfort. It allows you to focus on the moment. Screaming, just like exercise, laughing, and crying, releases endorphins, which affect the brain's limbic system, the areas that control pain and pleasure. Research has shown that when we vocalize our suffering, when we shout "OW!" loudly when something hurts, we actually are capable of enduring significantly more pain than if we grit our teeth and try to suppress our reaction. Scientists used to think that such outbursts were just a form of communication, a signal that you are in trouble, but now they understand them also as a form of natural pain relief.

The plane was so small, the ride incredibly bumpy. When the door opened, I was so frightened that my attention became very selective. I tried to focus on what the instructor said, but all I could hear over the whir of the propellers were the words "panic" and "scream."

"Got it!!"

I started screaming inside the plane, I kept screaming as we dove through clouds, I screamed the entire way down—forty seconds of free fall, which I realized almost immediately were the best forty seconds of my life since I had lost my husband. In this moment, I clearly understood that fear was synthesized in our brain just like happiness, that while we can't control what happens to us, we can control how we think about these things, even if it doesn't always feel this way.

This was the moment when I understood that the key to keeping John alive and in my life was facing the pain of remembering him, the pain of trying to embrace a ghost. Once I did that, once I faced my fear, I found him all around me. This was my last lesson: To love someone when they're gone just means holding them closer, keeping them in the part of your brain that feels like your heart.

Epilogue

A Holistic Theory of Love

Like millions of others, I faced the pandemic by myself, uncertain about what lay ahead. Would I be alone for the rest of my life? Would our social connections ever be the same? Would we get back to normal? The feeling of being cut off from the world was overwhelming. For many, that feeling was something new. But, for me, it was something I had been preparing for my whole life. And I was intrigued to watch as the rest of the world reacted to the same feelings of social isolation that had characterized my existence up to the point of meeting John, and that had fallen on me like an avalanche after his passing.

As soon as the pandemic hit, many of my fellow researchers in the social sciences scrambled to do experiments, trying to capture our unique frame of mind during this once-in-a-lifetime event. But I couldn't do experiments—the university buildings were closed, the laboratory shut down, the fMRI scanners powered off. I had to just sit back and watch the predictions from our research on loneliness play out in real life, in real time. I hoped that the pandemic, for all its challenges, would ultimately have a positive effect on people's social life, that it would force a huge society-wide mental reset, that people in our increasingly atomized and lonely world would learn how to connect with others and be more inclusive (even from a distance), learn why it is so important to prioritize relationships, learn that we can't take care of others unless we take care of ourselves.

By late March, Chicago was in a deep freeze and everything was locking down. I decided I needed a temporary change of scenery. I love Chicago—rain or shine—but I couldn't spend another crisis in the apartment. I needed greenery, I needed to be in nature, surrounded by trees and hope. Portland, Oregon, stuck out in my mind. John and I had visited that area in 2015. And we fantasized about one day settling there, buying a little house on Lake Oswego, only a few miles from the city. It took me a while but I finally found the house that we imagined, or one that looked the part, and rented it sight unseen.

At the beginning of the quarantine, airlines weren't flying. It wasn't safe to travel on trains or buses. But I had our car, hibernating in the garage of our building. I packed a few duffels and drove off with Bacio, our Chinese shar-pei, riding shotgun. I drove all the way from Chicago to Portland, three days straight, twelve hours each day. I took the longer, more mysterious, more beautiful northern route: Minneapolis, Fargo, Billings, Bozeman, Missoula, across the Spokane Valley, down the edge of Washington State, then along the Columbia River all the way to Portland. No one was on the road.

Every hotel I stayed in was entirely empty. At times, in the ice and snow of North Dakota and Montana, it felt less like driving and more like skating along the empty highways. I had snow and dirt caked thick on the car by the time I drove into town.

Once in Portland, I lived at first on packaged food: protein bars, cartons of soup. Then, once the lockdown was lifted, I started buying vegetables directly from nearby farms. I began each morning by running a few miles. Then I worked. I Skyped. I Zoomed. Like everyone else, I redefined what was normal. But there was something strange about this new normal: Hardly a week would go by without a journalist contacting me. First the *New York Times,* then the *Washington Post,* CNN, *Vogue, Women's Health, National Geographic*—they all wanted tips on surviving social isolation. They didn't really want to talk to Dr. Love, they wanted Dr. Loneliness. But since he was no longer around, they asked me to speak for John, to explain the research we had done together and my contributions to the development of clinical interventions to help reduce the effects of social isolation and facilitate social connection. Sometimes, reporters would actually confuse us, and whenever I got an email addressed to John Cacioppo, I would always smile—it felt like these "Dear John" emails were keeping him alive.

But, of course, there was a bitter note in receiving these emails: By being confronted time and time again with the reality that I could no longer forward them to John, I had to focus on all that I had personally lost. So to overcome the pain that I was reliving with every email, I had to intentionally look inward and gradually surface painful experiences and associate them with a more positive memory—a technique psychologists call cognitive-behavioral therapy (CBT), in which, according to neurologist Dr. Lisa Shulman, "people reduce their emotional charges by creating new mental associations." In my case, I remembered how John was happy to receive media requests or any emails with a question about science for that matter—he loved them. They gave him an opportunity to share his knowledge. So, with that in mind, every "Dear John" email started to make me think of his smile, and now I could see these emails in a more positive light.

In advising others by talking about our work on the science of loneliness, I tried to stay positive and objective. But I also found that to be authentic, to be true to myself, to connect with others, I had to self-disclose. I had to identify as one of the lonely masses, if only to tell people that I had made it back from the other side. Before the pandemic, I knew about the interpersonal health benefits of sharing positive news with others, but sharing *negative* experiences was new to me. The pandemic helped me better understand the virtues of collective sharing and how together we can improve our collective social capital (the inner strength we gain from our social connections). I understood now that talking about negative things was not the same as putting out negative energy. Emotions are merely emotions—neither positive nor negative. It's how we react to them that will determine whether they have a positive or negative impact on our health, our happiness, our longevity.

To give a sense of structure to my life, I woke up every day at four thirty, when it was still dark, serene, and peaceful outside. I meditated, I expressed gratitude for another day of life, I exercised. Then I took my laptop and settled in by a large window. Gazing at the stars, I felt that this pandemic life was not all that different from life on a space station, isolated yet hyperconnected. It was on one of these mornings that I opened my email to find a message from NASA. It was an invitation to give a virtual talk to the space agency, in conjunction with the National Institutes of Health, on the lonely brain. I wondered why on earth astronauts would be interested in my research. They are masters in the art of living under isolation.

They can spend a year alone in space, using positive thinking, structured routines, exercise, and their sense of mission to keep the effects of loneliness at bay. We should be learning from them.

That virtual event was unlike any other I had participated in. For confidentiality reasons, I couldn't see the attendees. I talked to a dark computer screen and answered fascinating questions from unidentified voices. I'm not sure what I taught the people at NASA, but the experience made me realize just how good a metaphor life in space was for our current reality. Like astronauts, we had to rewire our brain during the pandemic to stay close to our loved ones, even though they were sometimes very far away. From birthday celebrations to telemedicine, we had to move most of our social communications into the virtual realm.

I spent more time than usual stargazing during the pandemic. And one night, in the spring of 2021, I drove three hours south from Portland to a nature center and space observatory in the Sunriver meadow. The guide told me that I had impeccable timing. According to his calculations, in just a few minutes—at 11:22 P.M., to be precise—the International Space Station would pass overhead for twenty seconds and then vanish. The sun's rays reflecting off the space station made it bright enough to see with the naked eye, and it looked to me so much like a shooting star that I reflexively made a wish.

There was a full moon in the clear sky, almost zero light pollution, and I thought about the astronauts up there in space, surrounded by eighty-eight constellations, really just random dots until the human mind connected them together with the power of imagination. These were the same constellations that had made my curious mind thrive when I was a kid, that kept me company when I was a teenager, that showed me the way when I felt lost years later. The stars in that darkness made me think of a dear friend who would always remind me, "There is beauty in the struggle." In spite of all the challenges and dark moments we face, there is always a new way of seeing things, a new way to connect the dots. Sometimes we just have to remember to look up.

• • •

Looking back on it, I don't think my story is unique. I have met people of every shape and color who shared with me their own tales of love and heartbreak. And I always recognize myself in their joy as well as in their pain. Feelings like love and loneliness are universal—they cut across all categories, they include everyone. One of the surprising things about loneliness is that, unlike other chronic health risks, our social and economic status are not protective. Heartbreak comes for us all—chefs, athletes, nurses, doormen, physicists, poets, even pop stars.

Take Céline Dion, for example. Despite all the love songs she has made famous, most people don't know her real love story. Céline fell in love with her longtime business manager, René Angélil, when she was just becoming a star. He was a person she greatly admired, a person who supported her and guided her career from a young age. René had been twice divorced. René and Céline were separated by more than two decades in age. Céline's mother strongly opposed the marriage. For a time, Céline hid her true feelings. Yet they were too powerful, too pure, to suppress.

She and René decided to take a chance on love. The wedding was nationally broadcast on Canadian TV. Their love story played out in public. Céline held nothing back; she felt she had nothing to hide; she loved René totally. He was the only man she had ever been with, the only man she had ever kissed. They shared twenty-one happy years together. In 2016, after a long, brutal fight with esophageal cancer that forced him to spend the last years of his life eating through a tube, René died in Céline's arms, at age seventy-three. Two days later, Céline lost her beloved brother—also to cancer.

This is a woman who understands loss on a cellular level, who knows what it is to be embraced by love, who has likely experienced more than her fair share of loneliness. She never let René go, and even made a replica of her husband's hand, cast in bronze, which she grasps before she goes onstage every night. Almost from the moment Céline buried René, journalists wondered whether she could imagine falling in love for a second time. Six years after her husband's death, she startled them. "I am in love," she said. But she was also still single. *Alone, and in love.* "Love is not necessarily to marry again. When I see a rainbow, when I see a sunset, when I see a beautiful dance number—I'm in love. I go onstage every night because I *love* what I do."

What I want you to take from Céline's story, from mine, from the stories of so many others who have lost a loved one—and from all the lessons we've learned about the neuroscience of human connection—is that love is a much more expansive concept than we give it credit for. We must begin to view this phenomenon not as an isolated and ineffable emotion but as a cognitive and biological necessity, one that is measurable but ever changing, one that has the power to make us not only better partners but also better people. I began this book alone, and I'm ending it, well, alone. Yet by coming full circle, I believe that I have found the key to lasting love, both as a neuroscientist studying it in a laboratory and as a human being experiencing it in life. The key is to have an open mind. That is far, far easier said than done, but the process of opening the mind begins by understanding how it works.

That is exactly what you and I have tried to do in this book. Think about all that we know now. We know that love is a biological necessity. We know that social connections enabled the brain to evolve into the most powerful organ in the universe. We know that evolution also developed aversive signals—like loneliness and grief—to encourage us to take good care of our *social* body. We know that to be alone is not to be lonely. We know that love not only rewards us with a natural high but also satisfies a human need to expand our self. We know that such expansion requires us to dive into our inner space, to be honest, to be true to ourselves, to *reveal* who we are. We know, from the brain's perspective, that the love we feel for a person and a passion (like a sport or a career or a purpose in life) are very similar. We know that we cannot truly love without involving the mind, the heart, *and* the body. We know how tricky it is to hold on to love, how hard it is to let go of love, how damaging it is to lose love.

John used to talk about the fact that there is no word in English for "the opposite of loneliness." Like other biological needs—hunger, thirst—it's just something we lack an antonym for. But I'm starting to think that love—in the expansive way I now conceive of the term, based on my research and my experience—is the opposite of loneliness. Love is that feeling of social connectedness, of plenty, that John was searching for all these years. It is the thing that I feel surrounded by today. And, however your love story ends, I hope you are now even more inspired to find it for yourself.

With all my love.

Acknowledgments

Writing these last pages is a very humbling experience. In my deepest heart, I feel that every single person I have met during my life's journey has somehow inspired me or taught me a lesson in humanity. I am grateful beyond words to them all, to those I cite here (in no specific order), and to those I may not mention but who are nevertheless dear to me.

First, I would like to express my eternal gratitude to the love of my life, the one I met on a cold day in January and who warmed up my heart forever. You inspired me then and continue to inspire me every day in the most mysterious and beautiful ways. Your passion, your brilliance, your energy, your work ethic, your sophisticated creative thinking, your grace, and your unending love for people opened my heart and mind to a world I had never imagined possible. This world has an inner beauty and an inner truth that are simple and yet profound. It is a world where a smile can heal wounds of the mind; a world where joy, hope, and innovation are abundant; and a world where a meaningful life is a life connected to each other. Your physical absence still weighs heavily on me, but every day I keep you close in the part of my mind that feels like my heart.

I also wish to express my everlasting gratitude to all those we have lost. They inspired us and continue to inspire us to give the very best of ourselves every single day, to always be curious, to be kind to people and to never, ever take anything or anyone for granted.

Profound gratitude goes to John's and my entire family for their continuous support and love and to all those who have provided and expressed continuing support, care, and love to us. John's legacy will live on forever through his groundbreaking scientific theories and through all of us who were touched by him.

Special thanks to Laura Carstensen, Jack Rowe, and all the scientists who were with us in Paris, for turning a scientific meeting into a meeting of the hearts.

Endless gratitude to all the couples who inspired (and continue to inspire) me every day: my loving parents, the couples I cite in this book, the couples featured in the *New York Times*' "Modern Love" column, and my friends who are living happily ever after.

I deeply thank Stephen Heyman for his beautiful writing, his astute mind, and his patient, expert editing of the many drafts of this manuscript. By helping me translate my science and organize my poetic ideas, he has made a considerable contribution to the work as it is presented today.

Deep gratitude also to Steven Pinker, Elaine Hatfield, Richard Davidson, Giacomo Rizzolatti, Michael Gazzaniga, Scott Grafton, Jonathan Pevsner, Jean-René Duhamel, Richard Petty, and Ralph Adolphs for reading draft chapters or paragraphs of this book.

An immense debt of gratitude to my agent, Katinka Matson, for her brilliant mind and her steadfast support; to my editors at Flatiron, Megan Lynch and Meghan Houser, for their invaluable guidance and support in turning my life story and ideas into a book; and to everyone at Flatiron who made the publication of this book possible: Malati Chavali, Nancy Trypuc, Marlena Bittner, Erin Kibby, Christopher Smith, Kukuwa Ashun, Emily Walters, Vincent Stanley, Molly Bloom, Donna Noetzel, and Bob Miller.

I wish to also thank all the journalists, thinkers, and commentators whose inquiries and fascinating questions have challenged and pushed me over the past twenty years to expand the limits of my mind to new frontiers that I had never even imagined.

Deep thanks to all my mentors, colleagues, and friends in the science community, among whom are Elisabetta Làdavas, Alfonso Caramazza, Giacomo Rizzolatti, Michael Gazzaniga, Scott Grafton, Paolo Bartolomeo, Steve Cole, Stefano Cappa, Michael Posner, George Wolford, and Bruce McEwen. Throughout the years, they sparked my thinking, instilled in me the rigor of scientific methods, taught me the fundamentals of neuroscience, the principles of social and cognitive psychology, and complex mathematical approaches while still maintaining a joyful curiosity and sense of wonder in all scientific discoveries.

At the University of Chicago Pritzker School of Medicine, I am deeply grateful to all the nurses, staff, and physicians; Kenneth S. Polonsky, the president of the university's Medical Health System; Conrad Gilliam, the dean for basic science; Daniel Yohanna, the department chair of psychiatry and behavioral neuroscience; Bob Zimmer, the chancellor and former president of the University of Chicago; Shadi Bartsch-Zimmer and all my colleagues for providing me with a compassionate and invaluable intellectual environment.

Endless thanks to all the research participants for the generosity of their time and minds. I'm grateful to all my students, research assistants, and teaching assistants, who have been another source of inspiration—each impressed me with their passion, creativity, and dedication.

I'd also like to thank everyone at the universities where I studied or worked. They inspired me to push my limits and embark on the beautiful exploration of the mind. At Dartmouth College, I'd like to thank Leah Sommerville, Emily Cross, Antonia Hamilton, and all the other colleagues who also stayed long hours in the lab cranking out brain data before jumping back on our cross-country skis to head home. At the University Hospital of Geneva, I am deeply grateful to all the patients for their relentless inner strength and their inspiring inner peace, and all the nurses for their endless grace, their awe-inspiring calm, and their commitment to help patients through all sorts of difficult and emotional situations. Deep thanks also to all my mentors and colleagues on the neurology floor and in the departments of psychology and clinical neurosciences, especially Theodor Landis, Olaf Blanke, Margitta Seeck, Christoph Michel, Marie-Dominique Martory, Françoise Bernasconi, Jean-Marie Annoni, Fabienne Perren, Stephen Perrig, Pierre Mégevand, Patrik Vuilleumier, Armin Schnider, Francesco Bianchi-Demicheli, Paul Bischof, Dominique DeZiegler, Goran Lantz, and Claude-Alain Hauert. At the University of California in Santa Barbara, I'd like to thank all the professors and colleagues who inspired me by being pioneers in their respective fields, particularly Nancy Collins, Shelly Gable, Leda Cosmides, and Brenda Major. At Syracuse University, I wish to thank all my colleagues for their support, especially Amy Criss, William Hoyer, Larry Lewandowski, and Brian Martens. In Argentina, I thank all my colleagues at the Institute for Cognitive Neurology, particularly Facundo Manes, Blas Couto, and Agustin Ibanez, for their stimulating conversations, vast contributions to our scientific papers, and their inexorable passion for helping patients.

I am immensely grateful to all my college professors and my teachers from elementary school, especially Mr. and Mrs. Moreau-Gaudry and Mr. Roche, who taught me from a very young age how to share the gift of knowledge. And of course I owe a great debt of gratitude to my tennis coaches, who throughout my life have given me a deep sense of purpose and an everlasting love for sports.

Immense gratitude to all my West Coast friends, especially

Candice and Roger, Marylyn and Neil, John and Sharon, and Kim, Becky, Gucci, and Charlotte for their endless support at every hour of the day and night; and Sharon and Michael for their embracing hearts and minds. I would like to extend a limitless thank-you to all my friends in Oregon for constantly inspiring me.

Special thanks to all my wonderful friends, doormen, security guards, staff, and neighbors in Chicago. They know who they are. In particular, I am grateful beyond expressions to Fran and Marv, Jamie and Bruce, Lorna, Gail, Ann, Trish, Dawn and Tim, Shawn and Jeff, Patricia and John, Lorraine and Jon, Linda and John, Laura and John, Cathy and Craig, Debbie and Jim, Maureen and Sherwood, Mahtab and Sean, Angela, Elizabeth, Yolanda, Eric, Marvin, Cameron, Roy, Patrick, Emanuel, Arturo, Tom, Jerome, Jean-Claude, Joseph, and Jacob for their everlasting love for John, their profound inner strength, and their inspiring team spirit.

I am also profoundly grateful to my girlfriends, especially Fernanda, Leila, Nisa, Sandra, Nicole, Josée, Rosie, and Christiane, for their energizing enthusiasm, their perpetual encouragement, and their inspiring love for their respective partners and children. Their wicked sense of humor and their willingness to watch soccer games, attend US / Switzerland tennis matches at the Billie Jean King Cup, climb mountains, do yoga, or simply share their passion for knowledge has cheered me up over the past few years.

Profound gratitude to HRH the Crown Princess of Denmark for her endless support and heartfelt words and healing smiles during a time of my life when I needed them the most; and to Jane Persson, Helle Østergaard and all the teammates at the nonprofit Mary Foundation for their unconditional love and steadfast dedication to connecting people in the peaceful fight against loneliness.

Infinite thanks to my dog, Bacio, for taking me on walks with her and for her love and comfort.

Humble and respectful thanks to Coach for lighting a healing path, for teaching by example, for inspiring me to always look up as much as within, for instilling in me the everlasting joy of the pursuit of inner wisdom through sports, and for encouraging me to inspire others.

And finally, but not least, I would like to express my deep thanks to you—for reading these pages, for inspiring me to write this book and to share my story with you, even if that meant reliving it all. So, thank you for motivating me to dive deep within, for helping me to discover the silver lining in my story and to see beauty in everything . . . and everyone.

To be continued.

Notes

Introduction

2 **"Why do you dance?":** Graham Farmelo, *The Strangest Man: The Hidden Life of Paul Dirac, Mystic of the Atom* (New York: Basic Books, 2011), 187. See also Richard Gunderman, "The Life-Changing Love of One of the 20th Century's Greatest Physicists," *The Conversation*, December 9, 2015, accessed July 1, 2021, https://theconversation.com/the-life-changing-love-of-one-of-the-20th-centurys-greatest-physicists-51229.

2 **Margit Wigner . . . "you have made me human":** Farmelo, *The Strangest Man*, 284, 295–320.

3 **Half of adults in the United States are now single:** Nora Daly, "Single? So Are the Majority of U.S. Adults," pbs.org, September 11, 2014, accessed July 1, 2021, https://www.pbs.org/newshour/nation/single-youre-not-alone.

3 **According to a nationally representative survey carried out in 2020**: Center for Translational Neuroscience, "Home Alone: The Pandemic Is Overloading Single-Parent Families," *Medium*, November 11, 2020, accessed September 28, 2021, https://medium.com/rapid-ec-project/home-alone-the-pandemic-is-overloading-single-parent-families-c13d48d86f9e.

3 **a 2018 survey from Scotland:** J. H. McKendrick, L. A. Campbell, and W. Hesketh, "Social Isolation, Loneliness and Single Parents in Scotland," September 2018, accessed September 28, 2021, https://opfs.org.uk/wp-content/uploads/2020/02/1.-Briefing-One-180904_FINAL.pdf.

3 **from $1.69 to $3.08 billion:** David Curry, "Dating App Revenue and Usage Statistics (2021)," BusinessOfApps.com, March 10, 2021, accessed July 1, 2021, https://www.businessofapps.com/data/dating-app-market.

4 **used an online dating service:** Tom Morris, "Dating in 2021: Swiping Left on COVID-19," Gwi.com, March 2, 2021, accessed July 1, 2021, https://blog.gwi.com/chart-of-the-week/online-dating/.

4 **encouraging data:** John T. Cacioppo et al., "Marital Satisfaction and Break-ups Differ Across On-line and Off-line Meeting Venues," *Proceedings of the National Academy of Sciences* 110, no. 25 (2013): 10135–40.

4 **people prefer a limited range of choices:** Elena Reutskaja et al., "Choice Overload Reduces Neural Signatures of Choice Set Value in Dorsal Striatum and Anterior Cingulate Cortex," *Nature Human Behaviour* 2 (2018): 925–35.

5 **A Cambridge doctoral student:** "Lockdown Love: Pandemic Has Aged the Average Relationship Four Years," *Business Wire*, February 10, 2021, accessed July 1, 2021, https://www.businesswire.com/news/home/20210210005650/en/Lockdown-Love-Pandemic-Has-Aged-the-Average-Relationship-Four-Years.

5 **made their relationship stronger:** Lisa Bonos, "Our Romantic Relationships Are Actually Doing Well During the Pandemic, Study Finds," *Washington Post*, May 22, 2020, accessed July 1, 2021, https://www.washingtonpost.com/lifestyle/2020/05/22/marriage-relationships-coronavirus-arguments-sex-couples.

6 **distracted by their phone:** Pew Research Center, "Dating and Relationships in the Digital Age," May 2020, accessed July 1, 2021, https://www.pewresearch.org/internet/2020/05/08/dating-and-relationships-in-the-digital-age.

6 **classic headwinds facing couples:** Ellen S. Berscheid and Pamela C. Regan, *The Psychology of Interpersonal Relationships* (New York: Routledge, 2016), 429.

6 **not even on the dating market:** Pew Research Center, "Nearly Half of U.S. Adults Say Dating Has Gotten Harder for Most People in the Last 10 Years," August 2020, accessed July 1, 2021, https://www.pewresearch.org/social-trends/2020/08/20/nearly-half-of-u-s-adults-say-dating-has-gotten-harder-for-most-people-in-the-last-10-years.

6 **Japan is a particularly stark case:** Conrad Duncan, "Nearly Half of Japanese People Who Want to Get Married 'Unable to Find Suitable Partner,'" *The Independent*, June 19, 2019, accessed July 1, 2021, https://www.independent.co.uk/news/world/asia/japan-birth-rate-marriage-partner-cabinet-survey-a8966291.html.

6 **living without a spouse or a partner:** Richard Fry, "The Share of Americans Living Without a Partner Has Increased, Especially Among Young Adults," Pew Research Center, October 11, 2017, accessed July 1, 2021, https://www.pewresearch.org/fact-tank/2017/10/11/the-share-of-americans-living-without-a-partner-has-increased-especially-among-young-adults/.

6 **"Over and over, my undergraduates tell me":** Kate Julian, "The Sex Recession," *Atlantic*, December 2018, accessed July 1, 2021, https://www.theatlantic.com/magazine/archive/2018/12/the-sex-recession/573949/.

8 **"I have met in the streets":** Victor Hugo, *Les Misérables* (New York: Athenaeum Society, 1897), 312–13.

8 **"Love loves to love love":** James Joyce, *Ulysses* (Oxford: Oxford University Press, 1998), 319.

1. The Social Brain

12 **fifteen thousand words . . . "spreading wide his heart":** Desmond Sheridan, "The Heart, a Constant and Universal Metaphor," *European Heart Journal* 39, no. 37 (2018): 3407–9.

13 **"the heat and seething of the heart":** C. U. M. Smith, "Cardiocentric Neurophysiology: The Persistence of a Delusion," *Journal of the History of the Neurosciences* 22, no. 1 (2013): 6–13.

13 **Aristotle:** C. C. Gillispie, *Dictionary of Scientific Biography*, vol. 1 (New York: Charles Scribner's Sons, 1970).

13 **each organ constantly interacts:** Beatrice C. Lacey and John I. Lacey, "Two-Way Communication Between the Heart and the Brain: Significance of Time Within the Cardiac Cycle," *American Psychologist* 33, no. 2 (1978): 99. See also Rollin McCraty et al., "The Coherent Heart: Heart-Brain Interactions, Psychophysiological Coherence, and the Emergence of System-Wide Order," *Integral Review* 5 (2009): 10–115; Antoine Lutz et al., "BOLD Signal in Insula Is Differentially Related to Cardiac Function During Compassion Meditation in Experts vs. Novices," *Neuroimage* 47, no. 3 (2009): 1038–46.

14 **in *The Merchant of Venice*:** William Shakespeare, *The Merchant of Venice* (Shakespeare Navigators website), 3.2.63–64.

14 **Leonardo saw the brain:** Jonathan Pevsner, "Leonardo da Vinci's Contributions to Neuroscience," *Trends in Neurosciences* 25, no. 4 (2002): 217–20. See also Pevsner, "Leonardo da Vinci's Studies of the Brain," *The Lancet* 393 (2019): 1465–72.

14 **a "black box":** Sophie Fessl, "The Hidden Neuroscience of Leonardo da Vinci," Dana Foundation, September 23, 2019, accessed July 1, 2021, https://dana.org/article/the-hidden-neuroscience-of-leonardo-da-vinci.

16 **We have a lot of neurons—86 billion:** Frederico A. C. Azevedo et al., "Equal Numbers of Neuronal and Nonneuronal Cells Make the Human Brain an Isometrically Scaled-Up Primate Brain," *Journal of Comparative Neurology* 513, no. 5 (2009): 532–41.

16 **"only" 17 billion:** Michael S. Gazzaniga, "Who Is in Charge?," *BioScience* 61, no. 12 (2011): 937–38.

17 **more than a hundred thousand miles:** Lisbeth Marner et al., "Marked Loss of Myelinated Nerve Fibers in the Human Brain with Age," *Journal of Comparative Neurology* 462, no. 2 (2003): 144–52, https://doi.org/10.1002/cne.10714.

17 **one million gigabytes of information:** Paul Reber, "What Is the Memory Capacity of the Human Brain?," *Scientific American Mind*, May 1, 2010, accessed July 1, 2021, https://www.scientificamerican.com/article/what-is-the-memory-capacity/. See also Thomas Bartol Jr. et al., "Nanoconnectomic Upper Bound on the Variability of Synaptic Plasticity," *eLife* 4 (2015): https://elifesciences.org/articles/10778.

17 **4.7 billion books:** "Human Brain Can Store 4.7 Billion Books—Ten Times More Than Originally Thought," *Telegraph*, January 21, 2016, accessed July 1, 2021, https://www.telegraph.co.uk/news/science/science-news/12114150/Human-brain-can-store-4.7-billion-books-ten-times-more-than-originally-thought.html.

17 **a single twelve-watt lightbulb:** Sandra Aamodt and Sam Wang, *Welcome to Your Brain: Why You Lose Your Car Keys but Never Forget How to Drive and Other Puzzles of Everyday Behavior* (New York: Bloomsbury, 2009), 102. See also Ferris Jabr, "Does Thinking Really Hard Burn More Calories?," *Scientific American*, July 18, 2012, accessed July 1, 2021, https://www.scientificamerican.com/article/thinking-hard-calories/.

17 **millions of years ago:** Helen Fisher, *Anatomy of Love: A Natural History of Mating, Marriage, and Why We Stray*, rev. ed. (New York: W. W. Norton, 2017), 281.

19 **According to the social brain hypothesis:** Robin Dunbar, "The Social Brain Hypothesis," *Evolutionary Anthropology: Issues, News, and Reviews* 6, no. 5 (1998): 178–90.

19 **About seventy thousand years ago:** For a review on how humans (and their brains) evolved, see Yuval Noah Harari, *Sapiens: A Brief History of Humankind* (New York: Random House, 2014).

21 **the size of our individual social networks:** Rebecca Von Der Heide, Govinda Vyas, and Ingrid R. Olson, "The Social Network-Network: Size Is Predicted by Brain Structure and Function in the Amygdala and Paralimbic Regions," *Social Cognitive and Affective Neuroscience* 9, no. 12 (2014): 1962–72.

21 **If you raise a fish . . . A desert locust's brain:** Stephanie Cacioppo and John T. Cacioppo, *Introduction to Social Neuroscience* (Princeton, NJ: Princeton University Press, 2020), 77–83.

21 **some of the same brain regions:** Stephanie Cacioppo et al., "A Quantitative Meta-analysis of Functional Imaging Studies of Social Rejection," *Scientific Reports* 3, no. 1 (2013): 1–3.

21 **the anterior cingulate cortex:** Ibid., 206.

21 **less gray and white matter in key social areas:** Cacioppo and Cacioppo, *Introduction to Social Neuroscience*, 31–52.

22 **René Descartes, seeing water-powered automatons:** Michael S. Gazzaniga, *The Consciousness Instinct: Unraveling the Mystery of How the Brain Makes the Mind* (New York: Farrar, Straus and Giroux, 2018), 26–27.

22 **"the brain is a machine":** Matthew Cobb, *The Idea of the Brain* (New York: Basic Books, 2020), 1–2.

2. A Single Mind

27 **two out of three children:** Marjorie Taylor et al., "The Characteristics and Correlates of Fantasy in School-Age Children: Imaginary Companions, Impersonation, and Social Understanding," *Developmental Psychology* 40, no. 6 (2004): 1173–87.

27 **supramarginal gyrus:** Junyi Yang et al., "Only-Child and Non-Only-Child Exhibit Differences in Creativity and Agreeableness: Evidence from Behavioral and Anatomical Structural Studies," *Brain Imagining Behavior* 11, no. 2 (2017): 493–502.

27 **resist temptation:** B. J. Casey et al., "Behavioral and Neural Correlates of Delay of Gratification 40 Years Later," *Proceedings of the National Academy of Sciences* 108, no. 36 (2011): 14998–15003.

32 **less than two hundred milliseconds:** Stephanie Cacioppo, "Neuroimaging of Love in the Twenty-First Century," in *The New Psychology of Love*, ed. R. J. Sternberg and K. Sternberg (Cambridge, UK: Cambridge University Press, 2019), 357–68.

33 **rated their own photo:** Bruno Laeng, Oddrun Vermeer, and Unni Sulutvedt, "Is Beauty in the Face of the Beholder?," *PLoS One* 8, no. 7 (2013): e68395.

33 **one rather pungent study:** Claus Wedekind et al., "MHC-Dependent Preferences in Humans," *Proceedings of the Royal Society of London B (Biological Sciences)* 260, no. 1359 (1995): 245–49, https://royalsocietypublishing.org/doi/10.1098/rspb.1995.0087.

33 **well-fed female spiders:** Brain Moskalik and George W. Uetz, "Female Hunger State Affects Mate Choice of a Sexually Selected Trait in a Wolf Spider," *Animal Behaviour* 81, no. 4 (2011): 715–22.

34 **eating less when they are in love:** Elizabeth A. Lawson et al., "Oxytocin Reduces Caloric Intake in Men," *Obesity* 23, no. 5 (2015): 950–56.

3. Passion for Work

44 **I first encountered Huguette:** Olaf Blanke and Stephanie Ortigue, *Lignes de fuite: Vers une neuropsychologie de la peinture* (Lausanne: PPUR Presses Polytechniques, 2011), 113–43.

46 ***spatial hemineglect*:** Olaf Blanke, Stephanie Ortigue, and Theodor Landis, "Colour Neglect in an Artist," *The Lancet* 361, no. 9353 (2003): 264.

50 **the power of the brain to rewire itself:** For an overview of neuroplasticity, see Sharon Begley, *Change Your Mind, Change Your Brain: How a New Science Reveals Our Extraordinary Potential to Transform Ourselves* (New

York: Ballantine, 2007); and Norman Dodge, *The Brain That Changes Itself: Stories of Personal Triumph from the Frontiers of Brain Science* (New York: Penguin, 2007). Two other textbooks that cover this topic in detail are Eric R. Kandel, James H. Schwartz, and Thomas M. Jessell, *Principles of Neural Science* (New York: McGraw-Hill, 2012); and John T. Cacioppo, Laura Freberg, and Stephanie Cacioppo, *Discovering Psychology: The Science of Mind* (Boston: Cengage, 2021).

4. The Love Machine

53 **Before I came along:** These pioneering researchers who conducted some of the first neuro-imaging studies of romantic love include Andreas Bartels, Semir Zeki, Helen Fisher, Arthur Aron, Lucy Brown, Debra Mashek, Greg Strong, and Li Haifang.

56 **On my patent application:** Francesco Bianchi-Demicheli and Stephanie Ortigue, "System and Method for Detecting a Specific Cognitive-Emotional State in a Subject," U.S. Patent 8,535,060, issued September 17, 2013.

57 **a "low road":** Joseph LeDoux, *The Emotional Brain: The Mysterious Underpinnings of Emotional Life* (New York: Simon & Schuster, 1998), 161–64.

58 **a hundred milliseconds:** Raymond J. Dolan and Patrick Vuilleumier, "Amygdala Automaticity in Emotional Processing," *Annals of the New York Academy of Sciences* 985, no. 1 (2003): 348–55. See also Stephanie Ortigue et al., "Electrical Neuroimaging Reveals Early Generator Modulation to Emotional Words," *Neuroimage* 21, no. 4 (2004): 1242–51.

58 **Ralph Adolphs's patient S.M.:** Ralph Adolphs et al., "Impaired Recognition of Emotion in Facial Expressions Following Bilateral Damage to the Human Amygdala," *Nature* 372 (1994): 669–72. See also David Amaral and Ralph Adolphs, eds., *Living Without an Amygdala* (New York: Guilford Publications, 2016).

59 **I also conducted a "Love Machine" experiment:** Francesco Bianchi-Demicheli, Scott T. Grafton, and Stephanie Ortigue, "The Power of Love on the Human Brain," *Social Neuroscience* 1, no. 2 (2006): 90–103.

62 **while I scanned their brains using fMRI:** Stephanie Ortigue et al., "The Neural Basis of Love as a Subliminal Prime: An Event-Related Functional Magnetic Resonance Imaging Study," *Journal of Cognitive Neuroscience* 19, no. 7 (2007): 1218–30.

63 **Part of Einstein's genius:** Marian C. Diamond et al., "On the Brain of a Scientist: Albert Einstein," *Experimental Neurology* 88, no. 1 (1985): 198–204.

63 **other positive emotions:** Richard J. Davidson and William Irwin, "The Functional Neuroanatomy of Emotion and Affective Style," *Trends in Cognitive Sciences* 3, no. 1 (1999): 11–21; and Kristen A. Lindquist et al., "The Brain Basis of Emotion: A Meta-analytic Review," *Behavioral and Brain Sciences* 35, no. 3 (2012): 121.

64 **twelve specific brain regions:** Stephanie Cacioppo, "Neuroimaging of Love in the Twenty-First Century," in *The New Psychology of Love*, ed. R. J. Sternberg and K. Sternberg (Cambridge: Cambridge University Press, 2019): 332–44.

64 **compassion:** Emiliana R. Simon-Thomas, et al., "An fMRI Study of Caring vs Self-Focus During Induced Compassion and Pride," *Social Cognitive and Affective Neuroscience* 7, no. 6 (2012): 635–48. See also Matthieu Ricard, *Altruism: The Power of Compassion to Change Yourself and the World* (New York: Little, Brown, 2015).

65 **a cultural universal:** Elaine Hatfield and Richard L. Rapson, *Love and Sex: Cross-Cultural Perspectives* (Boston: Allyn & Bacon, 1996), 205.

5. Love in the Mirror

70 **Elaboration Likelihood Model:** Richard E. Petty and John T. Cacioppo, "The Elaboration Likelihood Model of Persuasion," in *Communication and Persuasion* (New York: Springer, 1986), 1–24.

70 **more affected by gut feelings:** Emeran A. Mayer, "Gut Feelings: The Emerging Biology of Gut-Brain Communication," *Nature Reviews Neuroscience* 12, no. 8 (2011): 453–66.

70 ***Thinking, fast and slow*:** Daniel Kahneman, *Thinking, Fast and Slow* (New York: Farrar, Straus and Giroux, 2011).

70 **"hub discipline":** John T. Cacioppo, Laura Freberg, and Stephanie Cacioppo, *Discovering Psychology: The Science of Mind* (Boston: Cengage, 2021), 70.

72 **their brain waves actually synchronize:** Alejandro Pérez, Manuel Carreiras, and Jon Andoni Duñabeitia, "Brain-to-Brain Entrainment: EEG Interbrain Synchronization While Speaking and Listening," *Scientific Reports* 7, Article 4190 (2017). See also Jing Jiang et al., "Neural Synchronization During Face-to-Face Communication," *Journal of Neuroscience* 32, no. 45 (2012): 16064–69.

74 **a simple "mirror game":** John T. Cacioppo and Stephanie Cacioppo, "Decoding the Invisible Forces of Social Connections," *Frontiers in Integrative Neuroscience* 6 (2012): 51. See also David Dignath et al., "Imitation of Action-Effects Increases Social Affiliation," *Psychological Research* 85 (2021): 1922–33, https://link.springer.com/article/10.1007/s00426-020-01378-1; and Rick B. van Baaren et al., "Mimicry and Prosocial Behavior," *Psychological Science* 15, no. 1 (2004): 71–74.

75 **Mirror neurons were first discovered:** Giacomo Rizzolatti and Corrado Sinigaglia, *Mirrors in the Brain: How Our Minds Share Actions and Emotions* (New York: Oxford University Press, 2008), 115.

76 **mirror neuron system could understand intentions:** Stephanie Ortigue et al., "Understanding Actions of Others: The Electrodynamics of the Left and Right Hemispheres: A High-Density EEG Neuroimaging

Study," *PLoS One* 5, no. 8 (2010): e12160. See also Stephanie Ortigue et al., "Spatio-Temporal Dynamics of Human Intention Understanding in Temporo-Parietal Cortex: A Combined EEG/fMRI Repetition Suppression Paradigm," *PLoS One* 4, no. 9 (2009): e6962.

77 **I invited experienced tennis players into the fMRI:** Stephanie Cacioppo et al., "Intention Understanding over T: A Neuroimaging Study on Shared Representations and Tennis Return Predictions," *Frontiers in Human Neuroscience* 8 (2014): 781.

6. When the Brain Swipes Right

83 **that is asexual:** Anthony F. Bogaert, "Asexuality: Prevalence and Associated Factors in a National Probability Sample," *Journal of Sex Research* 41, no. 3 (2004): 279–87. See also Esther D. Rothblum et al., "Asexual and Non-asexual Respondents from a U.S. Population-Based Study of Sexual Minorities," *Archives of Sexual Behavior* 49 (2020): 757–67.

83 **romantic preferences:** Dorothy Tennov, *Love and Limerence: The Experience of Being in Love* (Lanham, MD: Scarborough House, 1998), 74.

83 **attitudes around love and relationships:** G. Oscar Anderson, "Love, Actually: A National Survey of Adults 18+ on Love, Relationships, and Romance," AARP, November 2009, https://www.aarp.org/relationships/love-sex/info-11-2009/love_09.html.

84 **"everything: love, children" . . . "I feel no physical attraction":** Quentin Bell, *Virginia Woolf: A Biography* (New York: Harcourt Brace Jovanovich, 1974), 185.

84 **"you have given me":** Ibid., 226.

86 **"the rapture of being alive":** Joseph Campbell and Bill D. Moyers, *Joseph Campbell and the Power of Myth with Bill Moyers*, "Episode 2: The Message of Myth," New York: Mystic Fire Video, 2005.

86 **Using eye-tracking studies:** Mylene Bolmont, John T. Cacioppo, and Stephanie Cacioppo, "Love Is in the Gaze," *Psychological Science* 25, no. 9 (2014): 1748–56.

87 **awakens activity in a core brain:** Swethasri Dravida et al., "Joint Attention During Live Person-to-Person Contact Activates rTPJ, Including a Sub-Component Associated with Spontaneous Eye-to-Eye Contact," *Frontiers in Human Neuroscience* 14 (2020): 201.

87 **a landmark book:** Ellen Berscheid and Elaine Hatfield, *Interpersonal Attraction* (Reading, MA: Addison-Wesley, 1969).

87 **psychology of interpersonal relationships:** Ellen Berscheid and Pamela C. Regan, *The Psychology of Interpersonal Relationships* (New York: Routledge, 2016). See also Sarah A. Meyers and Ellen Berscheid, "The Language of Love: The Difference a Preposition Makes," *Personality and Social Psychology Bulletin* 23, no. 4 (1997): 347–62.

87 **its cultural evolution around the world:** Elaine Hatfield, Richard L. Rapson, and Jeanette Purvis, *What's Next in Love and Sex: Psychological and Cultural Perspectives* (New York: Oxford University Press, 2020); Elaine Hatfield and Richard L. Rapson, *Love and Sex: Cross-Cultural Perspectives* (Boston: Allyn & Bacon, 1996); and Elaine Hatfield and G. William Walster, *A New Look at Love* (Lanham, MD: University Press of America, 1985).

87 **separately or in concert:** Berscheid and Regan, *The Psychology of Interpersonal Relationships*. See also Cyrille Feybesse and Elaine Hatfield, "Passionate Love," in *The New Psychology of Love*, ed. R. J. Sternberg and K. Sternberg (Cambridge: Cambridge University Press, 2019), 183–207; Hatfield and Walster, *A New Look at Love*; and Lisa M. Diamond, "Emerging Perspectives on Distinctions Between Romantic Love and Sexual Desire," *Current Directions in Psychological Science* 13, no. 3 (2004): 116–19.

87 **the experience of *being in love*:** Berscheid and Regan, *The Psychology of Interpersonal Relationships*, 322–52, 373–74.

87 **groundbreaking theory:** Helen Fisher, *Anatomy of Love: A Natural History of Mating, Marriage, and Why We Stray*, rev. ed. (New York: W. W. Norton, 2017). See also Helen Fisher, "Lust, Attraction, and Attachment in Mammalian Reproduction," *Human Nature* 9, no. 1 (1998): 23–52; and Helen Fisher, "Anatomy of Love," Talks at Google, September 22, 2016, posted on December 7, 2016, https://www.youtube.com/watch?v=Wthc5hdzU1s.

89 **a key role in self-awareness:** Arthur D. Craig, "How Do You Feel—Now? The Anterior Insula and Human Awareness," *Nature Reviews Neuroscience* 10, no. 1 (2009). See also Richard J. Davidson and Sharon Begley, *The Emotional Life of Your Brain: How Its Unique Patterns Affect the Way You Think, Feel, and Live—and How You Can Change Them* (New York: Penguin, 2013), 318–24.

89 **desire wasn't lighting up the whole insula:** Stephanie Cacioppo, "Neuroimaging of Love in the Twenty-First Century," in *The New Psychology of Love*, ed. R. J. Sternberg and K. Sternberg (Cambridge: Cambridge University Press, 2019): 345–56.

90 **the dorsal part of the striatum:** Bernard W. Balleine, Mauricio R. Delgado, and Okihide Hikosaka, "The Role of the Dorsal Striatum in Reward and Decision-Making," *Journal of Neuroscience* 27, no. 31 (2007): 8161–65.

90 **a lesion confined to the anterior insula:** Stephanie Cacioppo et al., "Selective Decision-Making Deficit in Love Following Damage to the Anterior Insula," *Current Trends in Neurology* 7 (2013): 15.

92 **ideal romantic arrangement:** Emily A. Stone, Aaron T. Goetz, and Todd K. Shackelford, "Sex Differences and Similarities in Preferred Mating Arrangements," *Sexualities, Evolution & Gender* 7, no. 3 (2005): 269–76.

92 **a deficiency, if not a total absence, of physical desire:** Raymond C. Rosen, "Prevalence and Risk Factors of Sexual Dysfunction in Men and Women," *Current Psychiatry Reports* 2, no. 3 (2000): 189–95.

92 **seen as a cultural signifier of "marital bliss":** Sinikka Elliott and Debra Umberson, "The Performance of Desire: Gender and Sexual Negotiation in Long-Term Marriage," *Journal of Marriage and Family* 70, no. 2 (2008): 391–406.

93 **stress buffering:** India Morrison, "Keep Calm and Cuddle On: Social Touch as a Stress Buffer," *Adaptive Human Behavior and Physiology*, no. 2 (2016): 344–62.

7. We'll Always Have Paris

95 **long-distance relationships:** L. Crystal Jiang and Jeffrey T. Hancock, "Absence Makes the Communication Grow Fonder: Geographic Separation, Interpersonal Media, and Intimacy in Dating Relationships," *Journal of Communication* 63, no. 3 (2013): 556–77.

95 **even in elephants:** Ellen Williams et al., "Social Interactions in Zoo-Housed Elephants: Factors Affecting Social Relationships," *Animals* 9, no. 10 (2019): 747. See also "Elephant Emotions," *Nature*, October 14, 2008, accessed July 1, 2021, https://www.pbs.org/wnet/nature/unforgettable-elephants-elephant-emotions/5886/#.

98 **a decision-making game:** Robb B. Rutledge et al., "A Computational and Neural Model of Happiness," *Proceedings of the National Academy of Sciences* 111, no. 33 (2014): 12252–57. See also Bastien Blain and Robb B. Rutledge, "Momentary Subjective Well-Being Depends on Learning and Not Reward," *eLife* 9, e57977, https://elifesciences.org/articles/57977.

99 **they feel much less gratitude:** Giulia Zoppolat, Mariko L. Visserman, and Francesca Righetti, "A Nice Surprise: Sacrifice Expectations and Partner Appreciation in Romantic Relationships." *Journal of Social and Personal Relationships* 37, no. 2 (2020): 450–66.

100 **a mysterious brain area:** Sara M. Szczepanski and Robert T. Knight, "Insights into Human Behavior from Lesions to the Prefrontal Cortex," *Neuron* 83, no. 5 (2014): 1002–18.

100 **the regulation of emotions:** Richard J. Davidson and Sharon Begley, *The Emotional Life of Your Brain: How Its Unique Patterns Affect the Way You Think, Feel, and Live—and How You Can Change Them* (New York: Penguin, 2013), 43. See also Kevin N. Ochsner, Jennifer A. Silvers, and Jason T. Buhle, "Functional Imaging Studies of Emotion Regulation: A Synthetic Review and Evolving Model of the Cognitive Control of Emotion," *Annals of the New York Academy of Sciences* 1251 (2012): E1; and James J. Gross, ed., *Handbook of Emotion Regulation* (New York: Guilford, 2013).

101 **control and suppress urges:** Michael S. Gazzaniga, Richard B. Ivry, and G. R. Mangun, *Cognitive Neuroscience: The Biology of the Mind* (New York: W. W. Norton, 2014), 515–65.

101 **silver linings:** Ochsner, Silvers, and Buhle, "Functional Imaging Studies of Emotion Regulation."

101 **until about age twenty-five:** Mariam Arain et al., "Maturation of the Adolescent Brain," *Neuropsychiatric Disease and Treatment*, no. 9 (2013): 449.

101 **lack the ability to regulate our emotions:** Gazzaniga, Ivry, and Mangun, *Cognitive Neuroscience*, 468–73. See also Richard J. Davidson, Katherine M. Putnam, and Christine L. Larson, "Dysfunction in the Neural Circuitry of Emotion Regulation—a Possible Prelude to Violence," *Science* 289, no. 5479 (2000): 591–94; Antoine Bechara, Hanna Damasio, and Antonio R. Damasio, "Emotion, Decision Making and the Orbitofrontal Cortex," *Cerebral Cortex* 10, no. 3 (2000): 295–307; and Antoine Bechara, "The Role of Emotion in Decision-Making: Evidence from Neurological Patients with Orbitofrontal Damage," *Brain and Cognition* 55, no. 1 (2004): 30–40.

101 **manage psychological pain:** Wei-Yi Ong, Christian S. Stohler, and Deron R. Herr, "Role of the Prefrontal Cortex in Pain Processing," *Molecular Neurobiology* 56, no. 2 (2019): 1137–66.

101 **more like patients with a lesion:** Anat Perry et al., "The Role of the Orbitofrontal Cortex in Regulation of Interpersonal Space: Evidence from Frontal Lesion and Frontotemporal Dementia Patients," *Social Cognitive and Affective Neuroscience* 11, no. 12 (2016): 1894–901.

102 **His left orbitofrontal region:** John Darrell Van Horn et al., "Mapping Connectivity Damage in the Case of Phineas Gage," *PLoS One* 7, no. 5 (2012): e37454.

102 **"very energetic and persistent in executing all his plans":** John M. Harlow, "Passage of an Iron Rod Through the Head," *Boston Medical and Surgical Journal* 39, no. 20 (1848): 277.

102 **"no longer Gage":** Kieran O'Driscoll and John Paul Leach, "'No Longer Gage': An Iron Bar Through the Head: Early Observations of Personality Change After Injury to the Prefrontal Cortex," *BMJ* 317, no. 7174 (1998): 1673–74, doi:10.1136/bmj.317.7174.1673a.

102 **Lhermitte:** François Lhermitte, "Human Autonomy and the Frontal Lobes. Part II: Patient Behavior in Complex and Social Situations: The 'Environmental Dependency Syndrome,'" *Annals of Neurology* 19, no. 4 (1986): 336.

102 **faux pas:** Valerie E. Stone, Simon Baron-Cohen, and Robert T. Knight, "Frontal Lobe Contributions to Theory of Mind," *Journal of Cognitive Neuroscience* 10, no. 5 (1998): 640–56.

103 **rumination:** Aaron Kucyi et al., "Enhanced Medial Prefrontal-Default Mode Network Functional Connectivity in Chronic Pain and Its Association with Pain Rumination," *Journal of Neuroscience* 34, no. 11 (2014):

3969–75. See also Camille Piguet et al., "Neural Substrates of Rumination Tendency in Non-Depressed Individuals," *Biological Psychology* 103 (2014): 195–202.

103 **The prefrontal cortex on overdrive:** Davidson, Putnam, and Larson, "Dysfunction in the Neural Circuitry of Emotion Regulation": 591–94.

103 **cognitive enhancement:** Marc Palaus et al., "Cognitive Enhancement via Neuromodulation and Video Games: Synergistic Effects?," *Frontiers in Human Neuroscience* 14 (2020): 235. See also Gazzaniga, Ivry, and Mangun, *Cognitive Neuroscience*, 536–37.

103 **better mood:** Cortland J. Dahl, Christine J. Wilson-Mendenhall, and Richard J. Davidson, "The Plasticity of Well-Being: A Training-Based Framework for the Cultivation of Human Flourishing," *Proceedings of the National Academy of Sciences* 117, no. 51 (2020): 32197–206, https://doi.org/10.1073/pnas.2014859117. See also Jale Eldeleklioğlu, "Predictive Effects of Subjective Happiness, Forgiveness, and Rumination on Life Satisfaction," *Social Behavior and Personality* 43, no. 9 (2015): 1563–74; and Tamlin S. Conner and Paul J. Silvia, "Creative Days: A Daily Diary Study of Emotion, Personality, and Everyday Creativity," *Psychology of Aesthetics, Creativity, and the Arts* 9, no. 4 (2015): 463.

104 **the brains of Tibetan monks:** Davidson and Begley, *The Emotional Life of Your Brain*; Dahl, Wilson-Mendenhall, and Davidson, "The Plasticity of Well-Being," 32197–206. See also Nagesh Adluru et al., "BrainAGE and Regional Volumetric Analysis of a Buddhist Monk: A Longitudinal MRI Case Study," *Neurocase* 26, no. 2 (2020): 79–90; and Richard J. Davidson and Antoine Lutz, "Buddha's Brain: Neuroplasticity and Meditation," *IEEE Signal Processing Magazine* 25, no. 1 (2008): 176–74.

104 **over nine thousand hours of lifetime practice:** Tammi Kral et al., "Impact of Short- and Long-Term Mindfulness Meditation Training on Amygdala Reactivity to Emotional Stimuli," *Neuroimage* 181 (2018): 301–13.

104 **including the PFC:** Davidson and Begley, *The Emotional Life of Your Brain*, 744–841. See also Davidson and Lutz, "Buddha's Brain," 176–74; and Antoine Lutz et al., "Long-Term Meditators Self-Induce High-Amplitude Gamma Synchrony During Mental Practice," *Proceedings of the National Academy of Sciences* 101, no. 46 (2004): 16369–73.

104 **"convergence zone for thoughts and feelings":** "The Heart-Brain Connection: The Neuroscience of Social, Emotional, and Academic Learning," YouTube, https://www.youtube.com/watch?v=o9fVvsR-CqM.

104 **not only eminent monks:** Davidson and Begley, *The Emotional Life of Your Brain*, 754.

105 **specializes in positive emotions:** Ibid., 158–60.

105 **mindfulness apps:** An example of mindfulness apps can be found in: Dahl, Wilson-Mendenhall, and Davidson, "The Plasticity of Well-Being,"

32197–206. The app is entirely free and readers can learn about it by going to: tryhealthyminds.org.

105 **reduced neural activity in the parts of the PFC:** Gregory N. Bratman et al., "Nature Experience Reduces Rumination and Subgenual Prefrontal Cortex Activation," *Proceedings of the National Academy of Sciences* 112, no. 28 (2015): 8567–72. For more on the health benefits of nature, see Florence Williams, *The Nature Fix: Why Nature Makes Us Happier, Healthier, and More Creative* (New York: W. W. Norton, 2017).

8. Better Together

110 **"You know how well we fit together":** Sally Singer, "Ruben Toledo Remembers His Beloved Late Wife, Designer Isabel Toledo," *Vogue.com*, December 17, 2019, accessed July 1, 2021, https://www.vogue.com/article/isabel-toledo-memorial.

112 **facilitates creativity:** Allan Schore and Terry Marks-Tarlow, "How Love Opens Creativity, Play and the Arts Through Early Right Brain Development," in *Play and Creativity in Psychotherapy*, Norton Series on Interpersonal Neurobiology, ed. Terry Marks-Tarlow, Marion Solomon, and Daniel J. Siegel (New York: W. W. Norton, 2017), 64–91.

112 **synergy that leads to innovation:** Jen-Shou Yang and Ha Viet Hung, "Emotions as Constraining and Facilitating Factors for Creativity: Companionate Love and Anger," *Creativity and Innovation Management* 24, no. 2 (2015): 217–30; and Nel M. Mostert, "Diversity of the Mind as the Key to Successful Creativity at Unilever," *Creativity and Innovation Management* 16, no. 1 (2007): 93–100.

112 **oxytocin enhances creative performance:** Carsten K. W. De Dreu, Matthijs Baas, and Nathalie C. Boot, "Oxytocin Enables Novelty Seeking and Creative Performance Through Upregulated Approach: Evidence and Avenues for Future Research," *Wiley Interdisciplinary Reviews: Cognitive Science* 6, no. 5 (2015): 409–17.

112 **imagine taking a long walk:** Jens Förster, Kai Epstude, and Amina Özelsel, "Why Love Has Wings and Sex Has Not: How Reminders of Love and Sex Influence Creative and Analytic Thinking," *Personality and Social Psychology Bulletin* 35, no. 11 (2009): 1479–91.

112 **the more creative they consider themselves:** Kelly Campbell and James Kaufman, "Do You Pursue Your Heart or Your Art? Creativity, Personality, and Love," *Journal of Family Issues* 38, no. 3 (2017): 287–311.

112 **a series of experiments:** Stephanie Ortigue and Francesco Bianchi-Demicheli, "Why Is Your Spouse So Predictable? Connecting Mirror Neuron System and Self-Expansion Model of Love," *Medical Hypotheses* 71, no. 6 (2008): 941–44. See also Stephanie Ortigue et al., "Implicit Priming of Embodied Cognition on Human Motor Intention Understanding

in Dyads in Love," *Journal of Social and Personal Relationships* 27, no. 7 (2010): 1001–15; and Stephanie Cacioppo, Mylene Bolmont, and George Monteleone, "Spatio-Temporal Dynamics of the Mirror Neuron System During Social Intentions," *Social Neuroscience* 13, no. 6 (2018): 718–38.

113 **the mental states of strangers:** Rafael Wlodarski and Robin I. M. Dunbar, "The Effects of Romantic Love on Mentalizing Abilities," *Review of General Psychology* 18, no. 4 (2014): 313–21.

114 **what psychologists term *self-expansion*:** Arthur Aron and Elaine N. Aron, "Self-Expansion Motivation and Including Other in the Self," in *Handbook of Personal Relationships: Theory, Research and Interventions,* ed. Steve Duck (New York: John Wiley & Sons, 1997), 251–70.

114 **"The boundaries between you and not-you":** Barbara L. Fredrickson, *Love 2.0: Finding Happiness and Health in Moments of Connection* (New York: Penguin, 2013), 49.

115 **"When I'm not with you":** *Albert Einstein, Mileva Marić: The Love Letters,* ed. Jürgen Renn, Robert J. Schulmann, and Shawn Smith (Princeton, NJ: Princeton University Press, 1992), 23.

115 **out-of-body experiences:** Olaf Blanke et al., "Stimulating Illusory Own-Body Perceptions," *Nature* 419, no. 6904 (2002): 269–70.

116 **more activation we see in the angular gyrus:** Stephanie Ortigue et al., "The Neural Basis of Love as a Subliminal Prime: An Event-Related Functional Magnetic Resonance Imaging Study," *Journal of Cognitive Neuroscience* 19, no. 7 (2007): 1218–30.

9. In Sickness and in Health

127 **love literally makes us stronger:** Ellen S. Berscheid and Pamela C. Regan, *The Psychology of Interpersonal Relationships* (New York: Routledge, 2016), 31–62. See also Stephanie Cacioppo and John T. Cacioppo, *Introduction to Social Neuroscience* (Princeton, NJ: Princeton University Press, 2020), 21–53.

127 **Compared to single people:** Cacioppo and Cacioppo, *Social Neuroscience,* 22–23.

127 **2.5 times more likely to be alive:** Kathleen B. King and Harry T. Reis, "Marriage and Long-Term Survival After Coronary Artery Bypass Grafting," *Health Psychology* 31, no. 1 (2012): 55.

127 **couples' vital signs:** Kathleen B. King et al., "Social Support and Long-Term Recovery from Coronary Artery Surgery: Effects on Patients and Spouses," *Health Psychology* 12, no. 1 (1993): 56.

128 **tiny blister wounds:** Jean-Philippe Gouin and Janice K. Kiecolt-Glaser, "The Impact of Psychological Stress on Wound Healing: Methods and Mechanisms," *Critical Care Nursing Clinics of North America* 24, no. 2 (2012): 201–13.

128 **higher natural levels of oxytocin:** Jean-Philippe Gouin et al., "Marital Behavior, Oxytocin, Vasopressin, and Wound Healing," *Psychoneuroendocrinology* 35, no. 7 (2010): 1082–90.

129 **held their partner's hands:** James A. Coan, Hillary S. Schaefer, and Richard J. Davidson, "Lending a Hand: Social Regulation of the Neural Response to Threat," *Psychological Science* 17, no. 12 (2006): 1032–39.

129 **reduce the stress response:** A. Courtney DeVries, Erica R. Glasper, and Courtney E. Detillion, "Social Modulation of Stress Responses," *Physiology & Behavior* 79, no. 3 (2003): 399–407.

130 **that state of social deprivation:** For an overview on loneliness, see John T. Cacioppo and William Patrick, *Loneliness: Human Nature and the Need for Social Connection* (New York: W. W. Norton, 2008).

131 **being lonely increased the odds of an early death:** Julianne Holt-Lunstad et al., "Loneliness and Social Isolation as Risk Factors for Mortality: A Meta-Analytic Review," *Perspectives on Psychological Science* 10, no. 2 (2015): 227–37.

132 **sixty million people—report feeling so lonely:** Cacioppo and Patrick, *Loneliness*, 18.

132 **Chronic loneliness accelerates the aging process:** John T. Cacioppo and Stephanie Cacioppo, "The Growing Problem of Loneliness," *The Lancet* 391, no. 10119 (2018): 426. See also John T. Cacioppo and Stephanie Cacioppo, "Loneliness in the Modern Age: An Evolutionary Theory of Loneliness (ETL)," *Advances in Experimental Social Psychology* 58 (2018): 127–97; and Stephanie Cacioppo, John P. Capitanio, and John T. Cacioppo, "Toward a Neurology of Loneliness," *Psychological Bulletin* 140, no. 6 (2014): 1464.

135 **striking up a conversation with a stranger:** Nicholas Epley and Juliana Schroeder, "Mistakenly Seeking Solitude," *Journal of Experimental Psychology: General* 143, no. 5 (2014): 1980.

136 **try writing down five things:** Tara Lomas et al., "Gratitude Interventions," in *The Wiley Blackwell Handbook of Positive Psychological Interventions* (New York: John Wiley & Sons, 2014), 3–19.

136 **"micro-moments":** Barbara L. Fredrickson, *Love 2.0: Finding Happiness and Health in Moments of Connection* (New York: Penguin, 2013), 75.

137 ***Altruism*:** Matthieu Ricard, *Altruism: The Power of Compassion to Change Yourself and the World* (New York: Little, Brown, 2015).

137 **reduced feelings of loneliness among widows:** Dawn C. Carr et al., "Does Becoming a Volunteer Attenuate Loneliness Among Recently Widowed Older Adults?," *Journals of Gerontology: Series B* 73, no. 3 (2018): 501–10.

137 **"the benefits of solitude":** Micaela Rodriguez, Benjamin W. Bellet, and Richard J. McNally, "Reframing Time Spent Alone: Reappraisal Buffers

the Emotional Effects of Isolation," *Cognitive Therapy and Research* 44, no. 6 (2020): 1052–67.

138 **share good news:** Shelly L. Gable and Harry T. Reis, "Good News! Capitalizing on Positive Events in an Interpersonal Context," *Advances in Experimental Social Psychology* 42 (2010): 195–257. See also Ariela F. Pagani et al., "If You Shared My Happiness, You Are Part of Me: Capitalization and the Experience of Couple Identity," *Personality and Social Psychology Bulletin* 46, no. 2 (2020): 258–69; and Brett J. Peters, Harry T. Reis, and Shelly L. Gable, "Making the Good Even Better: A Review and Theoretical Model of Interpersonal Capitalization," *Social and Personality Psychology Compass* 12, no. 7 (2018): e12407.

10. The Test of Time

142 **socio-emotional selectivity:** Laura L. Carstensen, Derek M. Isaacowitz, and Susan T. Charles, "Taking Time Seriously: A Theory of Socioemotional Selectivity," *American Psychologist* 54, no. 3 (1999): 165. See also Wonjun Choi et al., "'We're a Family and That Gives Me Joy': Exploring Interpersonal Relationships in Older Women's Softball Using Socio-emotional Selectivity Theory," *Leisure Sciences* (2018): 1–18, doi:10.1080/01490400.2018.1499056.

142 **older people remembered far more:** Quinn Kennedy, Mara Mather, and Laura L. Carstensen, "The Role of Motivation in the Age-Related Positivity Effect in Autobiographical Memory," *Psychological Science* 15, no. 3 (2004): 208–14. See also Andrew E. Reed and Laura L. Carstensen, "The Theory Behind the Age-Related Positivity Effect," *Frontiers in Psychology* 3 (2012): 339; and Susan Turk Charles, Mara Mather, and Laura L. Carstensen, "Aging and Emotional Memory: The Forgettable Nature of Negative Images for Older Adults," *Journal of Experimental Psychology: General* 132, no. 2 (2003): 310.

143 **around 39 percent:** Belinda Luscombe, "The Divorce Rate Is Dropping: That May Not Actually Be Good News," *Time.com*, November 26, 2018, accessed July 1, 2021, https://time.com/5434949/divorce-rate-children-marriage-benefits/.

143 **breakup rates:** Roberto A. Ferdman, "How the Chance of Breaking Up Changes the Longer Your Relationship Lasts," *Washington Post*, March 18, 2016, accessed July 1, 2021, https://www.washingtonpost.com/news/wonk/wp/2016/03/18/how-the-likelihood-of-breaking-up-changes-as-time-goes-by/; and Michael J. Rosenfeld, "Couple Longevity in the Era of Same-Sex Marriage in the United States," *Journal of Marriage and Family* 76, no. 5 (2014): 905–18.

144 **lack of social reward is more decisive:** Yoobin Park et al., "Lack of Intimacy Prospectively Predicts Breakup," *Social Psychological and Personality Science* 12, no. 4 (2021): 442–51.

144 **breakups tend to spike:** Helen Fisher, "Evolution of Human Serial Pair Bonding," *American Journal of Physical Anthropology* 78, no. 3 (1989): 331–54.

145 **most newlyweds experienced:** Justin A. Lavner et al., "Personality Change Among Newlyweds: Patterns, Predictors, and Associations with Marital Satisfaction over Time," *Developmental Psychology* 54, no. 6 (2018): 1172.

145 **"OK. Which one?":** Hazel Markus and Paula Nurius, "Possible Selves," *American Psychologist* 41, no. 9 (1986): 954–69. See also William James, "The Consciousness of Self," chap. 10 in *The Principles of Psychology*, vol. 1 (New York: Henry Holt, 1890).

145 ***on paper*:** Michael S. Gazzaniga, Richard B. Ivry, and G. R. Mangun, *Cognitive Neuroscience: The Biology of the Mind* (New York: W. W. Norton, 2014), 573–78.

147 **anniversary issue of "Modern Love":** Stephen Heyman, "Hard-Wired for Love," *New York Times*, November 17, 2017, accessed July 1, 2021, https://www.nytimes.com/2017/11/08/style/modern-love-neuroscience.html.

148 **The experiment:** Arthur Aron et al., "The Experimental Generation of Interpersonal Closeness: A Procedure and Some Preliminary Findings," *Personality and Social Psychology Bulletin* 23, no. 4 (1997): 363–77.

149 **"Love didn't happen to us":** Mandy Len Catron, "To Fall in Love with Anyone, Do This," *New York Times*, January 11, 2015, accessed July 1, 2021, https://www.nytimes.com/2015/01/11/style/modern-love-to-fall-in-love-with-anyone-do-this.html.

149 **as self-disclosure increases:** Roy F. Baumeister, "Passion, Intimacy, and Time: Passionate Love as a Function of Change in Intimacy," *Personality and Social Psychology Review* 3, no. 1 (1999): 49–67; and Stephanie Cacioppo et al., "Social Neuroscience of Love," *Clinical Neuropsychiatry* 9, no. 1 (2012): 9–13.

149 **loneliness and a lack of self-disclosure:** Marcus Mund et al.,"Loneliness Is Associated with the Subjective Evaluation of but Not Daily Dynamics in Partner Relationships," *International Journal of Behavioral Development* (2020), doi:10.1177/0165025420951246. See also Marcus Mund, "The Stability and Change of Loneliness Across the Life Span: A Meta-Analysis of Longitudinal Studies," *Personality and Social Psychology Review* 24, no. 1 (2020): 24–52.

150 **activated the same regions:** Arif Najib et al., "Regional Brain Activity in Women Grieving a Romantic Relationship Breakup," *American Journal of Psychiatry* 161, no. 12 (2004): 2245–26.

151 **"I was not only heartbroken":** "Dessa: Can We Choose to Fall Out of Love?," filmed in June 2018 in Hong Kong, TED video, 11:31, https://www.ted.com/talks/dessa_can_we_choose_to_fall_out_of_love_feb_2019. For more about Dessa's adventures in neuroscience, see her

memoir, *My Own Devices: True Stories from the Road on Music, Science, and Senseless Love* (New York: Dutton, 2019).

11. Shipwrecked

160 **The memorial took place:** "Professor John T. Cacioppo Memorial," YouTube video, 56:17, posted by UChicago Social Sciences, May 7, 2018, https://www.youtube.com/watch?v=Fc2uEzTptxo.

12. How to Love a Ghost

163 **a longitudinal study of loneliness:** John T. Cacioppo, "Overcoming Isolation | AARP Foundation," YouTube video, 1:16, posted by AARP-Foundation, February 25, 2013, https://www.youtube.com/watch?v=xBWGdQ_lx_A.

164 **In the twenty-four-hour period after a loss:** Elizabeth Mostofsky et al., "Risk of Acute Myocardial Infarction After the Death of a Significant Person in One's Life: The Determinants of Myocardial Infarction Onset Study," *Circulation* 125, no. 3 (2012): 491–96.

164 **"broken heart syndrome":** Matthew N. Peters, Praveen George, and Anand M. Irimpen, "The Broken Heart Syndrome: Takotsubo Cardiomyopathy," *Trends in Cardiovascular Medicine* 25, no. 4 (2015): 351–57.

165 **40 percent higher risk:** C. Murray Parkes, Bernard Benjamin, and Roy G. Fitzgerald, "Broken Heart: A Statistical Study of Increased Mortality Among Widowers," *British Medical Journal* 1, no. 5646 (1969): 740–43.

165 **an elevated risk:** M. Katherine Shear, "Complicated Grief," *New England Journal of Medicine* 372, no. 2 (2015): 153–60.

165 **torments our brains:** Lisa M. Shulman, *Before and After Loss: A Neurologist's Perspective on Loss, Grief, and Our Brain* (Baltimore: Johns Hopkins University Press, 2018), 53–64.

165 **The brain's alarm center:** Manuel Fernández-Alcántara et al., "Increased Amygdala Activations During the Emotional Experience of Death-Related Pictures in Complicated Grief: An fMRI Study," *Journal of Clinical Medicine* 9, no. 3 (2020): 851.

165 **"regulating and planning" center:** Brian Arizmendi, Alfred W. Kaszniak, and Mary-Frances O'Connor, "Disrupted Prefrontal Activity During Emotion Processing in Complicated Grief: An fMRI Investigation," *NeuroImage* 124 (2016): 968–76.

167 **10 percent . . . "complicated grief":** Amy Paturel, "The Traumatic Loss of a Loved One Is Like Experiencing a Brain Injury," *Discover*, August 7, 2020, accessed July 20, 2021, https://www.discovermagazine.com/mind

/the-traumatic-loss-of-a-loved-one-is-like-experiencing-a-brain-injury. See also Shulman, *Before and After Loss*.

167 **the *nucleus accumbens*:** Mary-Frances O'Connor et al., "Craving Love? Enduring Grief Activates Brain's Reward Center," *Neuroimage* 42, no. 2 (2008): 969–72.

168 **more sensitive to the *anticipation*:** Brian Knutson et al., "Anticipation of Increasing Monetary Reward Selectively Recruits Nucleus Accumbens," *Journal of Neuroscience* 21, no. 16 (2001): RC159.

168 **likened it to a traumatic brain injury:** Shulman, *Before and After Loss*, 83–104.

168 **Eye-tracking studies:** Maarten C. Eisma et al., "Is Rumination After Bereavement Linked with Loss Avoidance? Evidence from Eye-Tracking," *PLoS One* 9, no. 8 (2014): e104980. For an overview of the science of grief, see Mary-Frances O'Connor, "Grief: A Brief History of Research on How Body, Mind, and Brain Adapt," *Psychosomatic Medicine* 81, no. 8 (2019): 731.

170 **to calm down the amygdala:** Richard J. Davidson and Sharon Begley, *The Emotional Life of Your Brain: How Its Unique Patterns Affect the Way You Think, Feel, and Live—and How You Can Change Them* (New York: Penguin, 2013), 285–95.

172 **"operant conditioning":** John T. Cacioppo, Laura Freberg, and Stephanie Cacioppo, *Discovering Psychology: The Science of Mind* (Boston: Cengage, 2021), 310.

173 **In 1941, they took a ferry:** James Gleick, *Genius: The Life and Science of Richard Feynman* (New York: Pantheon, 1992), 151.

174 **"idea-woman":** Richard P. Feynman to Arline Greenbaum, October 17, 1946, in *Perfectly Reasonable Deviations from the Beaten Track: The Letters of Richard P. Feynman*, ed. Michelle Feynman (New York: Basic Books, 2005), 68–69, https://lettersofnote.com/2012/02/15/i-love-my-wife-my-wife-is-dead, accessed July 20, 2021.

175 **when we shout "OW!":** Genevieve Swee and Annett Schirmer, "On the Importance of Being Vocal: Saying 'Ow' Improves Pain Tolerance," *Journal of Pain* 16, no. 4 (2015): 326–34.

EPILOGUE

179 **"people reduce their emotional charges":** Lisa M. Shulman, *Before and After Loss: A Neurologist's Perspective on Loss, Grief, and Our Brain* (Baltimore: Johns Hopkins University Press, 2018), 36.

180 **health benefits of sharing positive news:** Harry T. Reis et al., "Are You Happy for Me? How Sharing Positive Events with Others Provides Personal and Interpersonal Benefits," *Journal of Personality and Social Psychology* 99, no. 2 (2010): 311.

183 **"I am in love":** Catherine Thorbecke and Faryn Shiro, "3 Years After Her Husband's Death, Celine Dion Shares Advice to Overcome Loss: 'You Cannot Stop Living,'" *GoodMorningAmerica.com,* April 2, 2019, accessed July 20, 2021, https://www.goodmorningamerica.com/culture/story/years-husbands-death-celine-dion-shares-advice-overcome-62099061.

About the Author

Stephanie Cacioppo is one of the world's leading authorities on the neuroscience of human connections and emotions. Her work on the neurobiology of romantic love and loneliness has been published in top academic journals and covered by *The New York Times*, CNN, and *National Geographic*, among others.